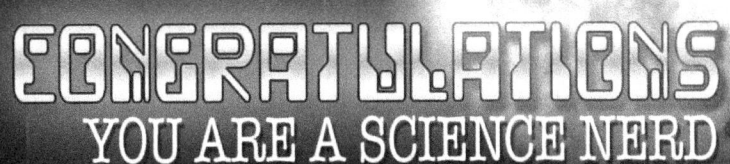

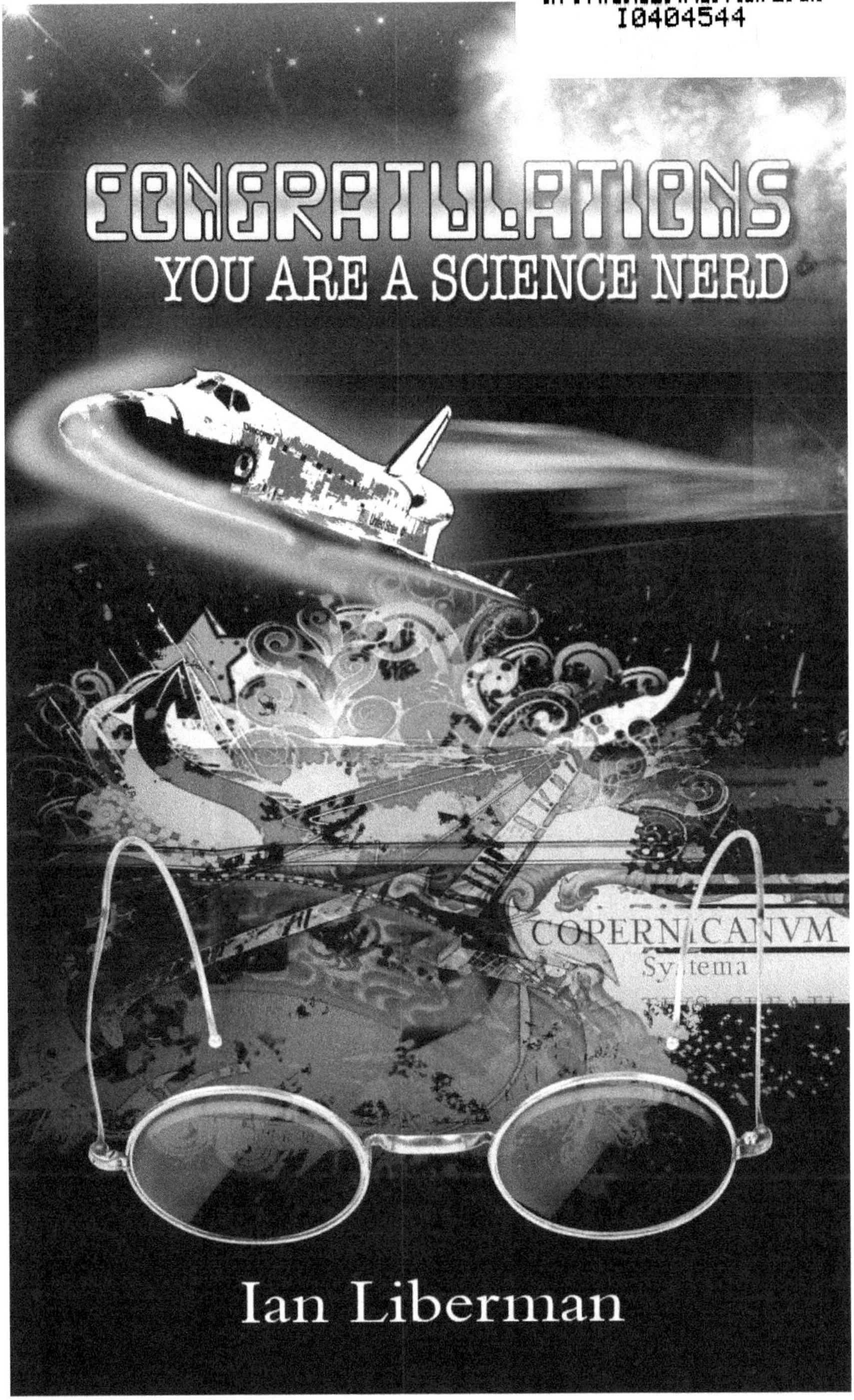
CONGRATULATIONS
YOU ARE A SCIENCE NERD
COPERNICANVM
Systema
Ian Liberman

Praise for Ian Liberman and Congratulations You are a Science Nerd

"In case you didn't get the memo, geeky is cool and nerdy is in. If anyone has doubts about that, Ian Liberman's brilliant book will prove the case page after page. Liberman's accessible and light touch with heavy science topics is the ideal formula for painless learning and memorable exploration. Young students and old science fans alike will love this exciting thrill-ride through a universe of wonders and unexpected turns."

—Guy P. Harrison, author of *Think: Why you should question everything*

"For all would-be Sheldons out there, this is the book for you. Ian Liberman combines his lengthy teaching experience with an avid interest in SF, comic books, and rock & roll to create a book that will appeal to nerds -- a complimentary term! -- of all ages."

David Darling, *Science writer and Astronomer*, *author The Rocket Man: And Other Extraordinary Characters in the History of Flight*

"An excellent resource for the public understanding of science! Ian Liberman leaves no stone unturned, covering a wide array of topics ranging from quantum physics and cosmology to the history of science and the societal implications of scientific awareness. I highly recommend, "Congratulations, You are a Science Nerd," to those looking to broaden their understanding of physics, cosmology, and modern science in the 21st century." **Tommy Rodriguez, a computational biologist**, *whose areas of interest is phylogenetics and evolutionary biology and is an advocate for the public understanding of science.*

My educational background has been primarily in the arts, but I have always been fascinated by the universe and how things work. Without a basic knowledge in the sciences, the vast amount of information available on the web and in books is often too complex to understand.

Ian Liberman's book explains, clearly and concisely, and with wonderful photographs and illustrations, everything you need to know to be on your way to becoming a Science Nerd.

The ebook format makes it easily accessible to everyone!

Gail Moss Baymiller
Senior Image Research Librarian
New York Public Library Picture Collection

I have read many books that try and present scientific ideas to the young and I must admit that most are anchored in pre-internet formats and written in a tone that fails to excite the imagination. Congratulations, You Are a Science Nerd understands how the minds of the new generation search for, take in and explore ideas and concepts. It presents science in a way that will excite the reader's imagination and provide meaning and understanding that will enhance his own human experience in a way that no religious belief can. In a sense, it is more than a book; it is a hub that connects the reader to the world of science. Excellent work!

Daniel Burnstein, physicist and creator of Quantum Geometric Dynamics.

Ian Liberman's does a masterful job at helping the average person understand truly complex ideas and theories in a very simple and clear way. The book answers many questions in the science world, from how the world was created to why Walter on Breaking Bad named himself Heisenberg. Mr Liberman uses humour, technology and effective visuals to engage his readers. After reading his book, I am proud to say that I am a science nerd!
 Carmelo Nanfara, Principal of Cedervale Community School (Toronto District School Board)

Contact:

Congratulations, you are a Science Nerd.

ianlib@rogers.com

Congratulations

You Are Science

Nerd

Copyright 2013 C by Ian Liberman

All Rights Reserved No part of this publication

Reproduced in any form or by any means, including

Scanning, photocopying, or otherwise without

prior written permission of the copyright holder.

Dedicated to Jill, Michael and Alana

Table of Contents

1. When and how was the earth created?
2. What does the universe consist of?
3. Is the Universe Fine Tuned?
4. What is the major theory of how the universe was created and why does it appear like magic?
5. What is the Big Bang Theory?
6. What is the Inflationary Theory of the Universe?
7. What is the Multiverse and what types are there?
8. Which comic books like to delve into the multiverse?
9. How would a parallel world or universe affect me?
10. Charles Robert Darwin (1809-1882)
11. Galileo Galilee (1564-1642)
12. Why was Isaac Newton's perception of the world, so important?
13. Let's review. What formula created this and who was one of the world's most brilliant scientists that created the formula?
14. What is Relativity?
15. What did Einstein really say about God and religion?
16. Who was the brilliant scientist of this century was gay, victimized and possibly committed suicide?
17. What was the Turing Test?
18. Why is Stephen Hawking`s accomplishments so amazing?
19. What is the force that is responsible the wear and tear on your joints?
20. What is the matter with Matter?
21. What are Dark Matter and Dark Energy and their relation to the Universe?

22. What is a vacuum?

23. What is another word for the Three Stooges? Entropy
24. What does Entropy have to do with Time?
25. What is a simple explanation of Quantum Physics?
26. Here is a picture of the atom and its parts.

27. What did the double slit experiment demonstrate and teach us about quantum mechanics or physics?

28. Schrödinger's Cat: How can a cat be alive and dead at the same time?
29. Who is Feynman and why was he so respected by his peers?
30. Why did Walter name himself Heisenberg in Breaking Bad?
 31. There are no ghosts or spirits.
32. What theory, in Quantum Physics, can give us teleportation, a perfect mail system and encryption for computers and quantum computers?
33. How does entanglement work to help us?

34. What is the Higgs Boson Particle and have we finally found it?

35. What is String Theory and why is it still a major contender for the Theory of Everything despite criticism?

36. What is the anti-world that resides next to us?

37. Who kidnapped Pluto and will we pay the ransom?

38. What telescopes, and technology, are searching the universe?

39. How does a quantum computer differ from a normal computer and what can it be used for?

40. Could our world or Universe be a hologram and are you really not kissing me?

41. What are holograms used for now? (See Picture display)

42. Can an equation predict life in space?

43. Is time travel possible?

44. What are your favorite movies or TV episodes using antimatter?

45. What are the most famous Particle Accelerators?

46. What is the scientific method and why is it important?

47. Nerd Guide to Famous Directors and their best movies and television programs. My top choices have photos.

48. Scientists and Individuals that you must look up and discover in the Nerd World.

Forward:

The real reason for writing this book is that I believe that North America is in the midst of an anti-science movement. This is hindering the development of new advances and ideas, in all areas of science, from physics to medicine.

Our school systems have minimal budgets for teaching curriculum and providing equipment for exploration and experimentation and as a result, our children are not experiencing the new and exciting ideas and concepts that are happening in science. Everybody can be a Nerd. Nerds use critical thinking skills. These skills can change your life in every way. The more Nerds there are, the more political and social change there will be in North America. It is a matter of scientific literacy. Science media stars like Michio Kaku, John Gribbin, Brian Greene, Andrew Zimmerman Jones, Lisa Randall, Neil de Grasse Tyson, Stephen Hawking, Leonard Mlodinow, Sean Carroll and others are trying to stimulate and educate the public about the really exciting and amazing things that are happening in the universe (and possibly other universes). The goal is to make all of us scientifically literate.

In order for our society to start producing more scientists, and help get Canada and the U.S. back to the forefront as leaders in science, we have to focus on the younger students in the junior grades, high schools and universities. We have to ignite their imaginations and build on their vision and resourcefulness. Brian Greene, Scientific American, Nov 9, 2011, provides good reasons why this book and others like it are needed; " One big thing is we often teach — not always — but we often teach we don't always teach the big ideas that make those details matter: how those details allow you to understand how the universe might have begun or where life might have come from, or how we're struggling to understand the nature of consciousness. Kids can get these big ideas, and in that way realize that science is not a bunch of facts and figures set inside a textbook, but rather science is a dynamic, living, breathing undertaking that can connect us to the universe in a far deeper way. And when Kids see science in that light, they're more excited to see the details, transforming the experience of learning science into something that's exciting. " We can say the same thing about adults.

http://blogs.scientificamerican.com/budding scientist/2011/11/09/ask-brian-greene-anything-really/

Having taught science for the best part of three decades and having created a Rock Music game that sold over a half of a million copies worldwide, I have come to know how to stimulate and relate to teenagers and young adults. Adults, in terms of science, need to comprehend and extrapolate the basics of physics and cosmology to get a proper worldview of how our world is changing based on scientific knowledge. This way they can confront pseudoscience and the misconceptions we have about theories like evolution and the universe. You do not need a certificate to become scientifically literate and to understand scientific concepts published by popular science writers. I can communicate with scientists because I have taught myself the fundamentals, in quantum physics, cosmology and biology. You can visit me on Facebook and learn from science postings and you can teach yourself with online courses. This book is part of the key to making you literate. My experience with lower grade levels will allow me to modify complex ideas to a simpler, plain and comprehensible level for you. I will show how these theories have been occurring in the pop culture of comics and science fiction. This book is for the young scientists at heart and the nerd who already sees the universe and our world from a very different pop cultural level. This book is also a book for the recreational reader at home and will appeal to both the Science Fiction Reader and the Comic Book buyer as well. When you finish this book, you will gain an understanding, of concepts that you find too difficult in the simplest science books constructed for the general public. This book will open up a new world for you because you will be able to pick up a book by Lawrence Krauss or John Gribbin or a magazine like Scientific American and understand the basic concepts of physics and cosmology.

You will find that many of the theories will be repeated, reinforced and simplified (just like here) so that you will have no difficulty understanding the major components and ideas. Are you ready to be exposed to the world of Science Fiction and Comics that was inspired by real physics and cosmology?

Let's dive into the cosmological questions of where we live and exist and with a guarantee that the answers are supernatural free, following only scientifically based theory. Let us start with a quickie about what came first, inflation or the Big Bang. Here are two articles. Which came first?
https://medium.com/starts-with-a-bang/25c5d719187b

http://www.lifesci.sussex.ac.uk/home/John_Gribbin/cosmo.htm

A: Inflation came before the Big Bang as far as most scientists are concerned. Learn more later.

Questions on the Universe:

What makes the world what it is? -The ATOM Firstly, atoms are VERY small. In a grain of sand (Silica) about a millimeter across, there are approximately 80,000,000,000,000,000,000 of them. The atom has an electron that is outside its nucleus. The nucleus consists of protons and neutrons. A nucleus has three "up" and three "down" quarks that are held together by gluons. Quarks bind the nucleus together. "Strong Force" holds together the gluons, the quarks, protons and neutrons. Basically, it holds the nucleus together and repels the electron preventing it from crashing into the proton. These are the building blocks of everything. Random is the key word in the behavior of the atom. Man can control the atom through science, but nothing else apart from the forces of "nature" does. This is important when we think of pseudoscience and religion.

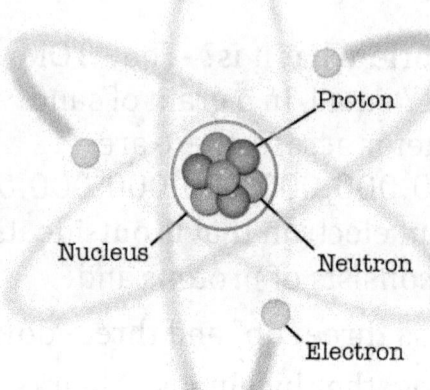
Copyright Steve Guest

1. When and how was the earth created?

The earth is the third planet from the sun, and at this point the only planet that we are aware of that has life. The earth is about the same age as the sun which is 4.7 billion years old. Earth is the only planet to have liquid water on its surface. In March 2014, it was discovered that Jupiter appears to have water vapor in the atmosphere. "The planet, named tau Boo b, orbits the nearby star tau Boötis and belongs to a class of exotic planets called "hot Jupiter's" that are not found in our solar system. A hot Jupiter is a massive extrasolar (outside our Solar system) planet that orbits very close to its parent star. Unlike our Jupiter, which is fairly cold and has an orbital period of about 12 years, tau Bootis orbits its star every 3.3 days and is heated to extreme temperatures by its proximity to the star." 1

1http://www.nrl.navy.mil/media/news-releases/2014/nrl-researchers-detect-water-around-a-hot-jupiter

There is no evidence that there is water on the surface of the planet but somewhere, in the history, there is a great possibility that there was water on the planet. We think Mars may have had water 3 to 4 billion years ago. Late in January 2013, evidence was found that showed that there may have been a substantial amount of water on Mars. "Strong geological evidence that Mars once had liquid water on its surface has been revealed this week. Scientists analyzing data sent back from NASA and ESA Mars orbiters say that the McLoughlin Crater, caused by a meteor impact millions of years ago, was probably once filled with water." (http://www.bbc.co.uk/programmes/p0139wlj)

    Mars seems to have 2 percent water, by weight, as found by the Mars Rover Curiosity in September, 2013. http://www.space.com/22949-mars-water-discovery-curiosity-rover.html . Liquid water covers 71% of the surface of the earth. The atmosphere is 77% nitrogen and 21% oxygen, with traces of other gases. The tilting of the earth towards the sun gives us the seasons. White clouds of water vapor hide much of the Earth's surface. The moon is our natural satellite with no atmosphere but many discerning features like craters, ridges, mountain ranges and lava plains. The moon revolves around us while the earth revolves around the sun. Temperatures on the moon can range from 100 degrees C to -170 degrees C. http://phys.org/news/2014-03-scientists-hot-jupiter.html#jCp

A hidden ocean has been found on Enceladus, a moon of Saturn, in the first week of April, 2014. Water jets from the southern polar region. This ocean may be under the surface of layers of rock. This is a very similar find to Europa.

The earth and the solar system formed when a cloud of gas and dust in space, was disturbed, possibly by a supernova. This explosion made waves in space which put pressure on the cloud of gas and dust. The pressure made the cloud start to collapse and as gravity pulled the gas and dust together, a solar "nebula" began to form. Finally, the cloud grew hotter and denser in the centre, while the disk of gas and dust that was surrounding it, though hot, was cooling at the edges. As the disk got thinner and thinner, particles began to clump together. Some clumps got bigger, as particles and small clumps stuck to them; they eventually became so large and massive that they formed planets or moons. It was near the centre of the cloud, where the planets like Earth formed. Just so you know, a "Supernova" is an exploding star, whose life has come to a spectacular end, while a solar nebula is a rotating cloud of gas and dust.

A
"Supernova "is an amazing sight. This explanation is

also is viable for how the solar system was created. Let's look at some images.

Earth

(NASA)

Supernova (NASA/ESA)

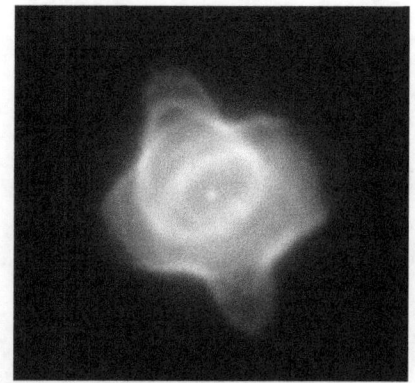

Solar Nebula

NASA/ESA

2. What does the universe consist of?

- The universe contains everything that exists; the Earth and all planets, all stars, all galaxies like our galaxy, the Milky Way, our solar system and other star systems like it. The universe was created 13.8 (Planck Satellite) billion years ago, after the Big Bang. New findings from NASA's

Kepler space observatory and the Proceedings of the National Academy of Sciences suggest at least twenty billion earth sized planets exist, as of November, 2013. However the first earth-sized planet, in the habitable zone, was found in April, 2014. http://www.themarysue.com/meet-kepler-186f-the-first-validated-earth-sized-planet-in-the-habitable-zone-of-a-red-dwarf/ There are hundreds of earth-like planets located in what we call the habitable zone, which means planets that orbit at a distance that is not too hot or too cold to support life and have liquid water.

- "Kepler has confirmed an additional 700+ exoplanets (planets outside our Solar system) orbiting 300 other stars.
- Ninety-five percent of these planets are smaller than Neptune.
- One hundred or so are roughly the size of Earth (!!!).
- Four of these new planets are in their star's habitable zone. (None validated).
- Like our solar system, these planets orbit in roughly the same plane." 2

---

2 http://www.slate.com/blogs/bad_astronomy/2014/02/26/exoplanets_new_technique_reveals_hundreds_more_alien_worlds.html

On March 23, 2013, the Planck Satellite provided us with an update of information on the universe by looking at CMB or Cosmic Microwave Background. To quote my own blog, "Cosmic Microwave Background is the remnants of the cooled down protons and electrons, created after the Big Bang, that have joined and become a microwave form of transparent light that has reached us and this light has now been detected by the Planck satellite. This technology has worked well."

What we have learned is that the universe is expanding at a slower rate than previously thought, and that the universe is 13.8 billion years old which is 100 million years older than the last measurement. Researchers estimate that the expansion rate of the Universe, known as Hubble's constant, is 67.15 plus or minus 1.2 km/second/mega parsec. Do not even bother to memorize this. This is more verification for the cosmologist; it endorses Pascal Jordan's early theories, and it supports Tryon, Linde and Guth's theories of Inflationary Cosmology. I will talk about these important individuals as we go along. A new major discovery about primordial gravitational waves is possibly in the works. The existence of these waves may substantiate the theory of Inflation. The important word there is "may". Many physicists like Sean Carroll, Edward Witten, Max Tegmark, John Gribbin, Lawrence Krauss, and especially Andrei Linde and Alan Guth who were involved in the modification and creation of Inflationary Cosmology, now feel vindicated. We still need physical verification of gravitational waves but it is looking good. This was provided John Gribbin`s journal or blog.

http://telescoper.wordpress.com/2014/05/14/that-bicep-rumour/

Inflationary cosmology primarily states that preceding the Big Bang, this expanding force, inflation, caused the universe to accelerate and continue for a short time. The BICEP2 paper states that inflation is "an early period of nearly exponential expansion which sets the initial conditions for the subsequent big bang." Ethan Sigel verifies this perspective. "So to recap, inflation is the thing that happens and that SETS UP THE BIG BANG. Based on what scientists have researched, it stretches the Universe flat, making it the same temperature in all directions, gets rid of any junk that was lying around beforehand, and creates the proper fluctuations on all scales — small and large — to grow into the stars, galaxies, and clusters that we have at present. If I were offering odds, I'd bet you 50-to-1 that inflation has it right. And that's all I've got for today!" http://scienceblogs.com/startswithabang/2011/03/16/what-are-the-odds-part-2-cosmi/   Here is Sean Carroll`s remarks to the BICEP2 findings. http://www.preposterousuniverse.com/blog/2014/03/16/gravitational-waves-in-the-cosmic-microwave-background/

We will talk about why the universe is still expanding or accelerating, later. Andrei Linde`s Theory of Chaotic Inflation is nicely described in the video by the World Science Festival (held in New York and a must see). http://www.worldsciencefestival.com/2011/04/bathrooms_and_bubble_universes/ This is a theory that talks about a process that continues to create pocket universes and a multiverse.

The new information and a major discovery have put the Inflation theory in a very positive light. Most scientists have supported the Big Bang Theory over other alternatives like the Cyclic and Loop Quantum Gravity. These alternatives had no Big Bang singularity or concept of quantum fluctuations. Stephen Hawking made a bet with Neil Turok, of the Perimeter Institute, that the Big Bang and Inflation were the main areas worth researching or investing in. At this point no money has exchanged hands. On Monday, March 7, 2014, there was a press conference held by Chou-Lin Kuo, of Stanford University, a participant of the South Pole BICEP2 telescope experiment. The experiment has been searching, analyzing and photographing CMB or Cosmic Microwave Background Fluctuations and Cosmic Extragalactic Polarization. BICEP means Background Imaging of Cosmic Extragalactic Polarization. The experiment found that CMB is slightly polarized and carry primordial gravitational waves. These are echoes of the beginning of the Big Bang at the billionth of a billionth, of a billionth of a millionth of a second after the Big Bang. The evidence finds support for Inflationary Cosmology and links quantum gravity during this period of rapid inflation. This is a Five Sigma discovery, having what is called a Gold Standard of certainty. Andrei Linde and Alan Guth were some of the major players in the theories of Inflationary Cosmology. Other ramifications may lead to information on dark energy and evidence for the multiverse through Quantum Field Theory. QFT leads to quantum fluctuations in the creation of the universe. The video on this discovery by Minute Physics is

excellent and the science is supervised by Sean Carroll. Check out
http://www.youtube.com/watch?v=4IlBNJbCzfk
and of course Sean`s Blog.
http://www.preposterousuniverse.com/blog/2014/03/19/a-great-time-for-reason-and-science/ .
John Gribbin`s writing on Inflation are also pretty amazing.
http://johngribbinscience.wordpress.com/2013/09/14/inflation-for-beginners-2/comment-page-1/

Scientists from the new Planck satellite project have also found the universe to be generally flat, and slightly lopsided. Planck has provided us with new information about the Cosmos. One of my favorite writers is Jennifer Ouellette discusses this.

http://blogs.scientificamerican.com/cocktail-party-physics/2013/03/21/walking-the-planck-our-lopsided-universe/

NASA/ESA (PUBLIC DOMAIN)

Finally the new percentages of dark matter and dark energy are here. "The rest of the universe appears to be made of a mysterious, invisible substance called **dark matter** and a force that repels gravity known as **dark energy**."
http://science.nationalgeographic.com/science/space/dark-matter/

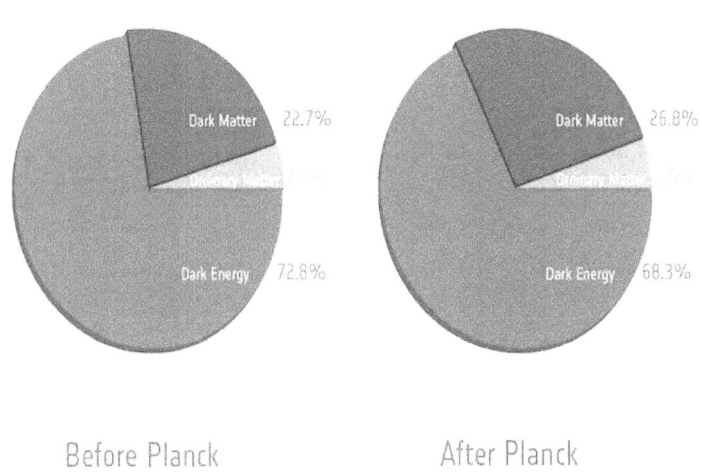

Before Planck   After Planck

Copyright: ESA and the Planck Collaboration
Here is an official report from NASA.
http://www.jpl.nasa.gov/news/news.php?release=2013-109#1

Most of the universe is a mess of uninhabited, desolate, explosive, dying matter. The majority of scientists believe that there is no fine-tuning of life in the universe and we are here from a random accident. This is because the universe has been around long enough for this creation of earth, to happen and that the odds were that the chemical soup that needed to bind together to form life, happened on a random basis, and could occur again and again and that we are the proof that the conditions were right, here on earth. In another universe, there may not be the same chemicals needed to create life, because the laws of physics could be different. Also there may be other forms of life that may not be carbon based.

3. Is the universe fine tuned?

Astronomer Brandon Carter developed the Anthropic Theory to answer that question. We will consider the Weak Anthropic Principal which states that if our universe was not hospitable to life, we would not be here to be thinking about this concept. The universe, of course, is not hospitable to carbon based life. So it seems that this is a universe of probabilities, not fine tuning. Victor Stenger, physicist, states this, from his paper called, "Is the Universe Fine Tuned for us?" "If indeed, multiple universes exist, then we are simply in that particular universe of all the logically consistent possibilities that had the properties needed to produce us."

http://www.colorado.edu/philosophy/vstenger/Cosmo/FineTune.pdf

Planck indicated that the universe is, flat and partially lopsided, with zero energy and this leads to our next hypothesis about how the universe was created. This is the theory that suggests a self sustaining universe, without the necessity of God, and the theory is backed many cosmologists of the National Academy of Science. As Victor Stenger states, "Only 7 percent of the members of the NAS, the elite of American scientists, believe in a personal god." This was a poll from 1998. Sean Carroll wrote an excellent article about why virtually all cosmologists are atheists.

http://preposterousuniverse.com/writings/nd-paper/

A universe is created as a quantum fluctuation or a virtual particle in a vacuum or field of nothingness and then propelled as from a singularity (small point) by inflation. Again, this is a major theory called Inflationary Theory. Others include Loop-Gravity and the Aeons' Theory.

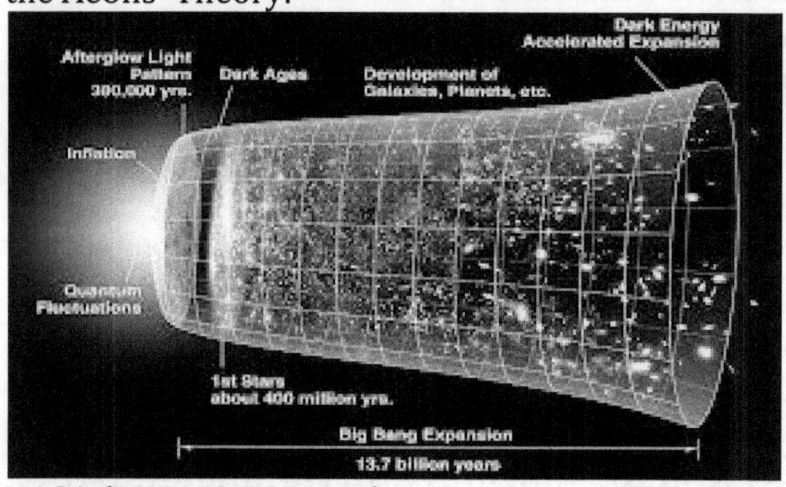

NASA (PUBLIC DOMAIN)

4. What is the major theory of how the universe was created and why does it appear like magic?

Let us look at the ultimate magic trick from the Cosmos, plucking universes out of nothing. Magicians pluck something from nothing all the time. Well, maybe not universes, but rabbits and coins. However, it is never supernatural or magic. Most of the top scientists in the world, such as, Leonard Mlodinow, Stephen Hawking and Lawrence Krauss, Edward Whitten and

Leonard Susskind thought that the evidence indicates that our own earth may have come from an area of nothingness or void where quantum fluctuations appear as a universe or universes. The theory originates from Quantum Field Theory, based on a subatomic level, where virtual particles (particles that are not real) are created in fields such as an electromagnetic field and the virtual particles can be an anti-photon, photon or an electron. Here is the Particle Zoo and the Atom Image:

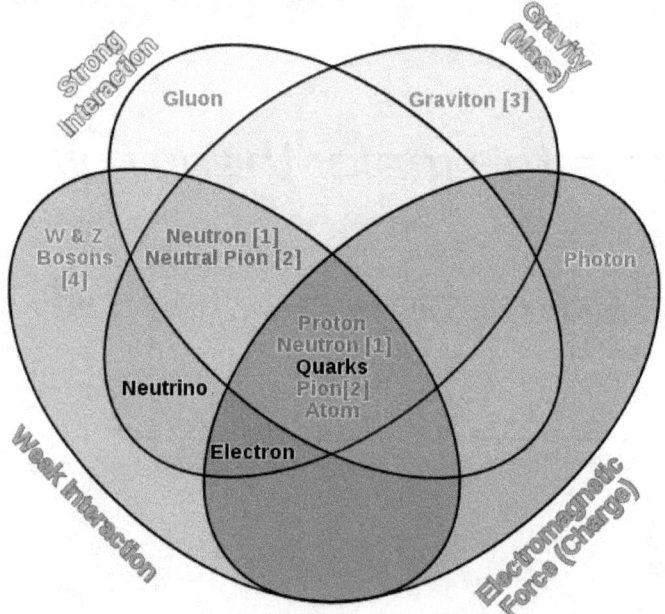

Particles_by_fundamental_interactions.svg Georg-Johann

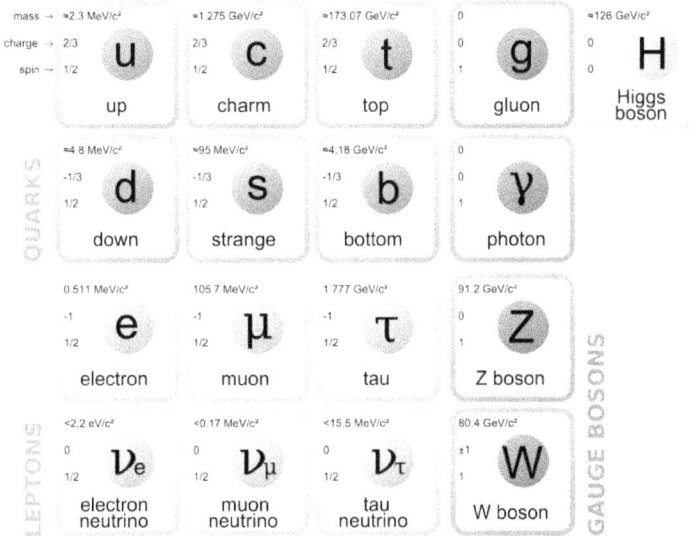

Author MissMJ and Standard_Model_of_Elementary_Particles.svg

In A Universe from Nothing, Lawrence Krauss describes what happens in the area in space where there is nothing. "In quantum gravity, universes can, and indeed always will, spontaneously appear from nothing. Such universes need not be empty, but can have matter and radiation in them, as long as the total energy, including the negative energy associated with gravity balancing the positive energy of matter is zero."[3] Quantum Gravity and inflation contribute to the expansion of the universe before the big bang. Inflation expands those quantum fluctuations to create a universe that is flat and has almost zero energy. That has been suggested by NASA using CMBR or again what we call Cosmic Microwave Background Radiation

---

[3] Lawrence N. Krauss, A Universe From Nothing (New York, NY, Free Press, 2012) Pg. 169.

remnants. For many scientists, inflation came before and set up the Big Bang, as suggested by Ethan Siegel and Max Tegmark.
http://scienceblogs.com/startswithabang/2011/02/02/can-you-get-something-for-noth/ and http://www.nytimes.com/2014/04/12/opinion/big-bang-to-little-swoosh.html?_r=0

   In Quantum Field Theory, on the subatomic level, anti-protons, electrons, and protons are popping in and out of existence all the time in fields. This is the concept applied to an area of nothingness before the universe is created in space. In QFT, virtual particles pop in and out of fields like magnetic fields or electric fields or electromagnetic fields. In the 1930`s, Paul Dirac talked about the Dirac Sea and Quantum Field Theory, where virtual particles or vacuum fluctuations that did not exist for long periods, popped in and out of existence from fields or vacuums on a subatomic level. Dirac thought of the idea of antiparticles, when he formulated the theory of antimatter. He saw the anti-world of particles as a whole new world. He theorized about a second version of matter and particles that are the opposite of what existed now. This was called "antimatter".

   "In 1928, physicist Paul Dirac was creating the quantum theory of the electron, when he realized that the equation only worked if you allowed these extra particles — identical to electrons, but with opposite charge — to exist. Just four years later, the first positrons were discovered and named by Carl D. Anderson while he was analyzing cosmic rays."[4] The electrons anti particle was called the positron while the proton was the antiproton and the neutron was the anti-neutron. Take this concept and apply it into space, and we see the spontaneous creation of universes as quantum vacuum fluctuations or quantum fluctuations from a vacuum of nothingness. In the vacuum or field of nothingness where virtually almost zero energy, it is speculated that inflation takes the quantum fluctuations to produce the singularity of the Big Bang that starts the expansion of the universe. I have also mentioned, in the book, that the singularity state, before the Big Bang is controversial. This zero energy stops the virtual universe from disappearing like a virtual particle would on a subatomic level. Fluctuations continue to make the galaxies and solar systems. Also, it is hypothesized; inflation continues to make pocket universes as one universe expands. It is called Eternal or Chaotic Inflation.

Fields are everywhere. If we do not look at the fields (present are usually energy or force fields), there are vibrations in the fields but when looking at them or measuring them, we see virtual particles or antiparticles moving in and out of them. This visual analogy was inspired by Sean Carroll. http://www.symmetrymagazine.org/article/july-2013/real-talk-everything-is-made-of-fields . In these fields or vacuums that are in space, we find the reasons why we have close to zero energy universes. Gravity is negative energy and everything else related

---

[4] A.Zimmerman and D.Robbins, String Theory for Dummies(Hoboken,NJ:Wiley Publishing ,2010) pg127

to matter is positive which produces flat zero energy universes. "The matter in the universe is made out of positive energy. However, the matter is all attracting itself by gravity. Two pieces of matter that are close to each other have less energy than the same two pieces a long way apart, because you have to expend energy to separate them against the gravitational force that is pulling them together. Thus, in a sense, the gravitational field has negative energy. In the case of the universe that is approximately uniform in space, one can show that this negative gravitational energy exactly cancels the positive energy represented by the matter. So the total energy of the universe is zero." [5]

Spontaneous universe creation theories came from other scientists, going all the way back to the thirties, with Pascal Jordan and Edward Tryon. Here is Edward Tryon`s paper. It is fascinating to read and compare to the two most recent books on spontaneous creation to Tryon`s paper on the creation of the universe. Lawrence Krauss and Stephen Hawking`s books are worth a look. Tryon`s theory of the balancing or cancelling of negative gravitational energy with the positive energy of the matter in the universe to produce a zero energy universe has been carried through and modified by many scientists, as part of inflationary universe model. Here is his paper for those interested.

http://pdfcast.org/download/tryton1973-vac-fluc.pdf

---

[5] Stephen Hawking, A Brief History of Time (New York, NY: Random House, 1996) Pg. 166.

Let us look at NASA's perspective on Quantum fluctuations and vacuums in space. As NASA states, "The longer WMAP observes, the more it reveals about how our universe grew from microscopic quantum fluctuations to the vast expanses of stars and galaxies we see today."
http://www.nasa.gov/home/hqnews/2006/mar/HQ_06097_first_trillionth_WMAP.html

The quantum vacuum fluctuations create potential universes, through a process that resemble carbonated bubbles that float out of a can of ginger ale. Bubbles are coming in every direction and position, bumping into each other, some dying, and some living, just as they do in the Quantum world, with all the possible histories of particle creation. This is where Quantum Field Theory is involved. Since there is zero energy, in the vacuum, the bubbles survive longer.

As I have discussed, these quantum fluctuations were the beginning of creation for stars, galaxies, Milky Ways, solar systems, the Big Bang and finally human beings in our universe. Many scientists believe that space is infinite, so this is happening all the time. Cosmologists like Professor Guth and Linde (who initiated the inflationary universe theory) speculated that some universes eventually will be producing a small percentage of planets that have life like ours because the universe creation process may be an infinite process. As Kepler Space Telescope data reported a possible 10 to 20 billion in the habitable zone in November, 2013. Remember we are not talking about intelligent life. The odds are against that.

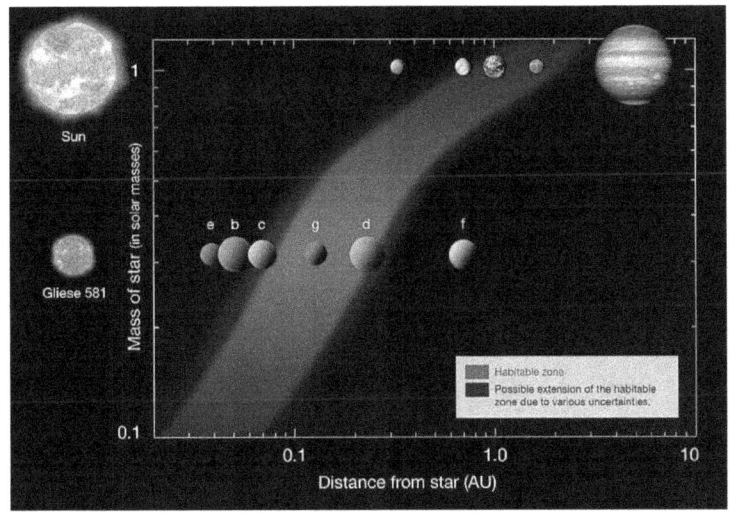

Source http://www.eso.org/gallery/d/89840-4/phot-15b-09-fullres.jpg

Author ESO

If anyone asks you how was the universe was formed, just say quantum fluctuations which produced a flat, lopsided universe out of nothing; spontaneous creation. Another excellent book to research and learn, about this idea, is A Universe from Nothing by Lawrence Krauss or In Search of the Multiverse by John Gribbin. That sums it up. I have repeated this major theory a few times, so you will understand this basic concept involves a self perpetuating universe with "no supernatural shenanigans" as Dr. Krauss has said in many lectures and books. http://www.radcliffe.harvard.edu/event/2013-lawrence-krauss-lecture

Let us talk about when and how the Casimir Effect takes place. From my own blog quotes. "The nothing can be both a field and labeled a vacuum. Space is filled as with electromagnetic fields and waves." To understand the Casimir Effect, one first has to understand something about a vacuum in space as it is viewed in quantum field theory. Far from being empty, modern physics assumes that a vacuum is full of fluctuating electromagnetic waves that can never be completely eliminated, like an ocean with waves that are always present and can never be stopped. These waves come in all possible wavelengths, and their presence implies that the empty space contains a certain amount of energy; energy that we can tap.

The Casimir Effect works on a subatomic level in space and is based on Quantum Field Theory. "The effect illustrates that a vacuum is also a field which allows the spontaneous creations of universes in space and particles on a subatomic level. In this case, we are looking at an

electromagnetic field. The mirror experiment also predicted vacuum energy and vacuum fluctuations which scientists applied to quantum field theory making it quantum vacuum fluctuations."
http://www.scientificamerican.com/article.cfm?id=what-is-the-casimir-effect .

Another good description is from Science Daily. "The Casimir force arises from effects of the vacuum (empty space). According to quantum mechanics, the vacuum contains fleeting electromagnetic waves, in turn consisting of electric and magnetic fields."
http://www.sciencedaily.com/releases/2007/02/070220144759.htm

Also let's talk about the "nothing," again as Jerry Seinfeld would say. I give this quick little speech for those who do not understand.

I also state this from a post on my blog, "The nothing is a quantum vacuum or field and the universe is basically the creation of a quantum vacuum fluctuation in the virtually empty vacuum or field. It has its basis from quantum field theory. The universe balances out to zero energy as gravity is negative energy and matter is positive, and they basically cancel each other out. Zero energy Universe. That is why the universe was conceived, in this vacuum to be flat and why on a subatomic level virtual particles created in fields or vacuums disappear right and others pop up right away. However, in space, the zero energy allows many of these virtual particles or universes to stay and not disappear. This is how the universe comes from nothing." I have outlined this before in the book but reinforcement is the prime strategy for learning.

NASA (PUBLIC DOMAIN)

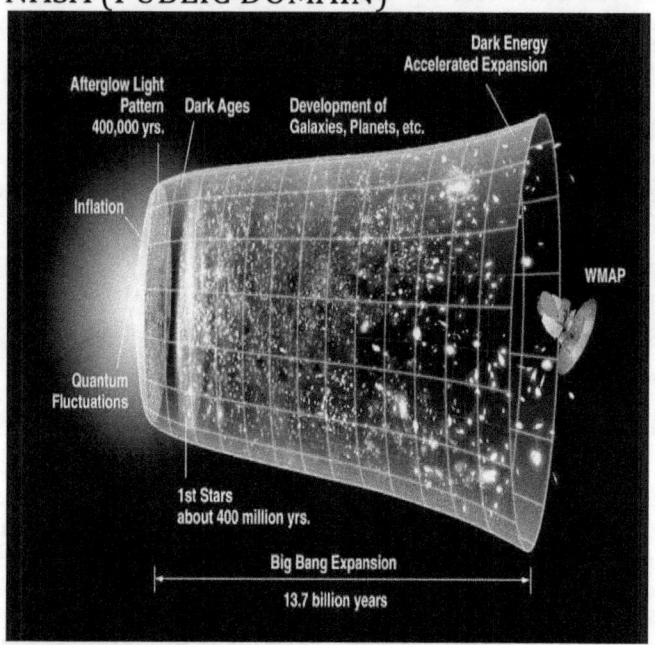

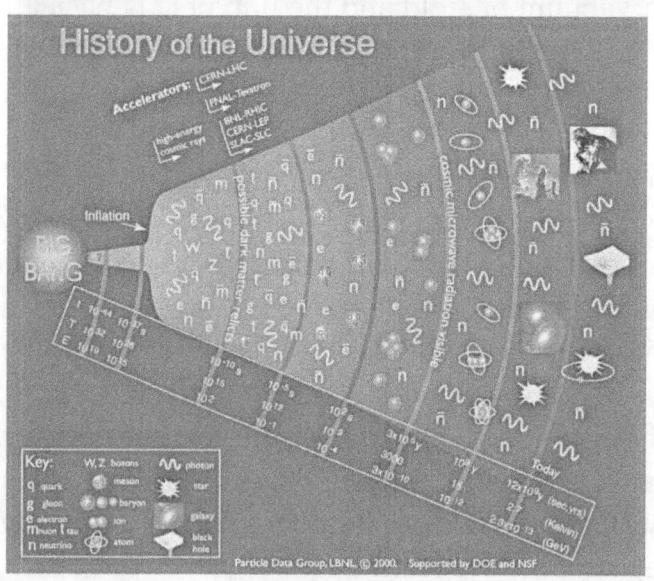

5. What is the Big Bang Theory?

There was a gravitational singularity which precipitated the expansion of the universe and this was called the Big Bang Theory. (Easier to tell your friends this version)

Created in the new space-time environment out of nothing, over 13.8 billion years ago and started with inflation. A inflationary fluctuation from a spec (singularity) became hot and dense and voila the Big

Bang; fluctuations continued to create the universe with galaxies, while continuing to expand with the process of inflation. Inflation petered out, and there is speculation that dark energy is still maintaining the expansion.

This work is licensed under the Creative Commons Attribution-ShareAlike 3.0 License.

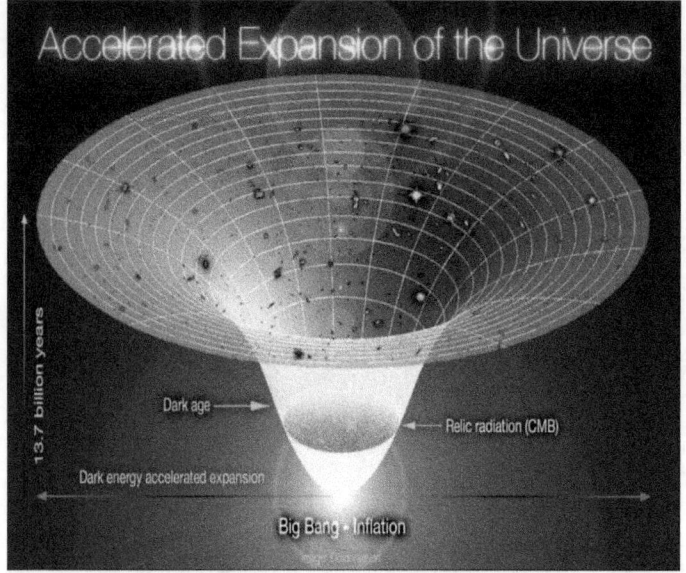

This licensing tag was added to this file as part of the GFDL licensing update. *This file is licensed under the Creative Commons Attribution ShareAlike 2.5 Netherlands licenceAttribution: Design Alex Mittelmann, Coldcreation*

The Big Bang Theory is also an American sitcom created by Chuck Lorre; it premiered on CBS on September 24, 2007. It revolves around physicists, relationships, timely and accurate theories in physics.

Author: Mihailodalj

6. What is the Inflationary Theory of the Universe?

More reinforcement of this theory, as we are doing now, contributes to being a functional nerd. The Inflation Theory was developed by Alan Guth, Andrei Linde, Paul Steinhardt, and Andy Albrecht. In our life, when we think of inflation when we go to buy a pizza and the price is more than it was last year. We eat the double, cheese pepperoni pizza and guzzle the down the diet drink and wake up and get up go to the scale. Look at our stomach. We see inflation. With inflation, items, get bigger and expand. Inflation happens all around us. The above scientists thought that at the beginning of the universe, random quantum vacuum inflationary fluctuations, out of nothing, caused our universe to expand and possibly be created. Many speculated that this phenomenon could be manifesting on a multiple level causing the Multiverse and that many universes would be created as pocket universes. However, there is still a consensus including Alexander Vilenkin, Andrei Linde, Alan Guth, Lawrence Krauss, Leonard Mlodinow, Sean Carroll, Brian Greene, and Stephen Hawking, that the universe is and has been rapidly expanding since the big bang, defying the gravitational pull. It is thought that Dark Energy may be a factor in the present expansion which was the cosmological constant in Einstein`s theories. He decided to reject that explanation. It may have been his biggest mistake or not.

7. What is the Multiverse and what types are there?

Let`s quickly talk about Erwin Schrodinger, who was known for his quantum theories relating to superposition and particle behaviour. His theories included an equation for wave function probability. Superposition is a particle that is actually in all possible states simultaneously, until we measure it. He wanted to know if this could work in the world today and not just in the subatomic world. So he developed Schrodinger`s Cat. From Whatis.com , http://whatis.techtarget.com/definition/Schrodingers-cat "Here is Schrödinger's (theoretical) experiment: We place a living cat into a steel chamber, along with a device containing a vial of hydrocyanic acid. There is, in the chamber, very small amount of hydrocyanic acid, radioactive material. If even a single atom of the substance decays during the test period, a relay mechanism will trip a hammer, which will, in turn, break the vial and kill the cat."
http://whatis.techtarget.com/definition/Schrodingers-cat

The observer cannot know whether an atom of the substance has decayed, and consequently, cannot know whether the vial has been broken, the hydrocyanic acid released, and the cat killed. It is a fifty/fifty chance that this quantum event will have the cat dead or alive.

Since we cannot know, according to physics, whether the cat is both dead and alive, the cat is in two states at one time; this is what is called a superposition of states. It is only when we break open the box and learn the condition of the cat that the superposition is lost, and the cat becomes one or the other (dead or alive) .This is the collapse of the wave function. (Copenhagen solution)

From the Encyclopedia Britannica: "wave function, in quantum mechanics, is the variable quantity that mathematically describes the **wave characteristics** of a particle. The value of the wave **function** of a particle, at a given point of space and time, is related to the likelihood of the particle's being there at the time." We call this wave probability.
http://www.britannica.com/EBchecked/topic/637845/wave-function

"The observation or measurement itself affects an outcome, so that the outcome as such does not exist unless the measurement is made." (That is; there is no single outcome unless it is observed.)
http://whatis.techtarget.com/definition/Schrodingers-cat and Watch the video http://www.youtube.com/watch?v=7SjFJlmg2Z8

Remember the wave function of the particle represents the probability of where the particle could be when it is in superposition. It is a really a probability wave.

Vectorisation of **Multiverse —level II.GIF** (by **K1234567890y**), by Lokal_Profil

Firstly let's examine the different theories that allow Parallel Worlds or Universes to exist. Two of the major physicists that discuss and have expertise, in these areas, are Max Tegmark and Brian Greene. Here are the universes and parallel Earths proposed by those two involved. Those recommended by Max Tegmark are:

1) Since Space is infinite; there are probable regions beyond our cosmic horizon that involve duplicate earths and other universes. This is called the Quilted Universe.
2) The inflationary bubble universes were created as almost zero energy vacuums with quantum random post-inflation fluctuations. These flat universes manifest with every possible history of universes produced. Stephen Hawking, Guth and Linde are just some of the supporters and contributors to this theory which is the most popular of

all theories. These and other scientists have modified this concept with different approaches, but the basic premise is the same. E.g. Lawrence Krauss

3) The Many Worlds or the Parallel Universe or Parallel World Concept by Hugh Everett. It is an alternative to the collapse of the wave function or the Copenhagen Effect. The Copenhagen solution takes place when we measure or observe a particle or object existing in a superposition state, or when particles are in two or more states at once. Once the measurement occurs we are left with one stable state. (Copenhagen Version)

The Many Worlds Theory insists that there are worlds created for each state and when the box is opened there is no collapse of the wave function. Instead of Quantum Decoherence (the reason for the collapse of the wave function which ends the superposition of particles or objects), coherence is present and the cat is alive on some worlds and dead in many different worlds. This is the kind of event we love in science fiction and comic books but is actually taken quite seriously by most scientists. When applied to Schrödinger's cat, the cat can be alive here and dead on a parallel world. The MWT can occur on a subatomic level, as well. Then we measure the spin or polarization of an electron or photon, which is embodying more than one defined characteristic or state, at the same time (This is called superposition) . The states can include spin up or spin down or horizontal and vertical polarization of the particle. With MWT, we find that rather than a wave function collapse happening, a new parallel world is created with one state present on the parallel world and the other at the original location. See how this happens in Schrödinger's Cat.

With the Copenhagen solution, when we open the box, the observation causes the cat to become dead or alive. With MWT it branches out into parallel worlds. While the box was closed it was theoretically in a state of superposition, being dead and alive. In MWT, Schrodinger's Cat is alive in one world and dead in another world in the Many Worlds Theory.

4) Mathematical Universes. These universes are located in several extra dimensions that may be discovered in many of the new theories like String Theory.

They have their own set of physical laws and are all around us. Max Tegmark has a new book called The Mathematical Universe that describes this scenario.

Photo:Arenamontanus **Anders Sandberg Flickr.**

Some other types of multiverse, included in Brian Greene's Book, called The Elegant Universe include,
1) Simulated Multiverse: A universe which is probably created by a quantum computer; something that the famous Seth Lloyd would not disagree. He wrote the book Programming the Universe which states that the universe has the characteristics of a quantum computer and may be one.

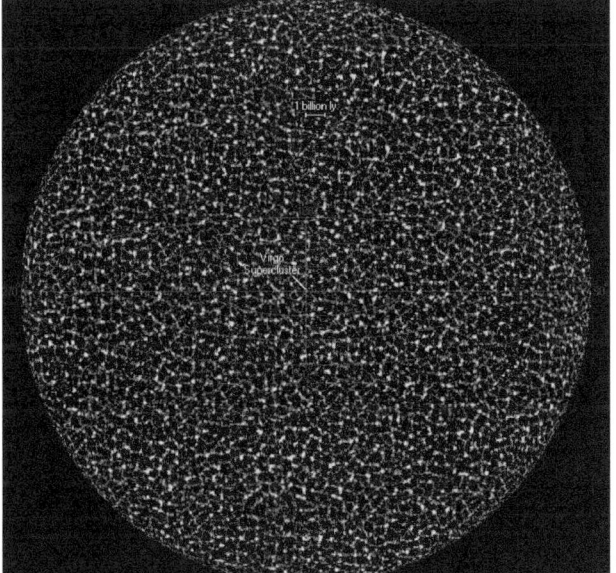
Richard Powell

2) Holographic Multiverse: This is a mirrored universe based the Holographic Principal where our universe is mirrored in the form of a holograph based on the information of a duplicated universe, decoded on a flat 2d surface, at the edge of our universe.

Daniele Clozza - www.atlantidea.org **Holographic U.**

3) Landscape Multiverse: More Inflationary Bubble Universes created from nothing with different physical laws.

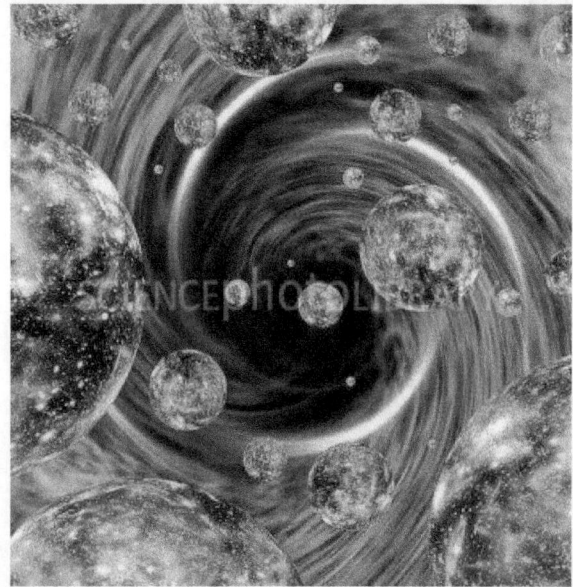
Bubble Universes.

4) There is the Brane Multiverse from String Theory: Our universe is just connected to one of the 3-dimensional branes or membranes (branes), while other branes may

have other numbers of dimensions and could have whole other universes on them. In String Theory, there are branes, which resemble membranes and the theory contains 8 or more dimensions. For more info, see Andrew Zimmerman`s blog.
http://physics.about.com/od/physicsatod/g/brane.htm
5) Cyclic Multiverse or Big Bounce: Branes just colliding and bumping into each other forming new universes. This is called an Ekpyrotic universe.

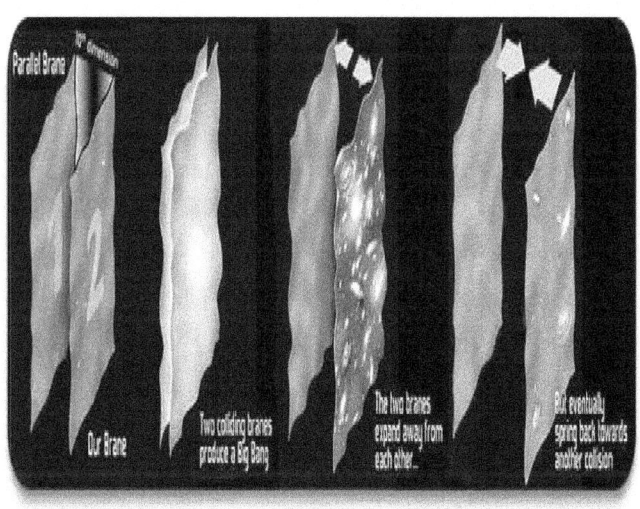

http://imageshack.us/a/img84/4082/collision.jpg

Copyright © 2010 Jean-Luc Lehners.

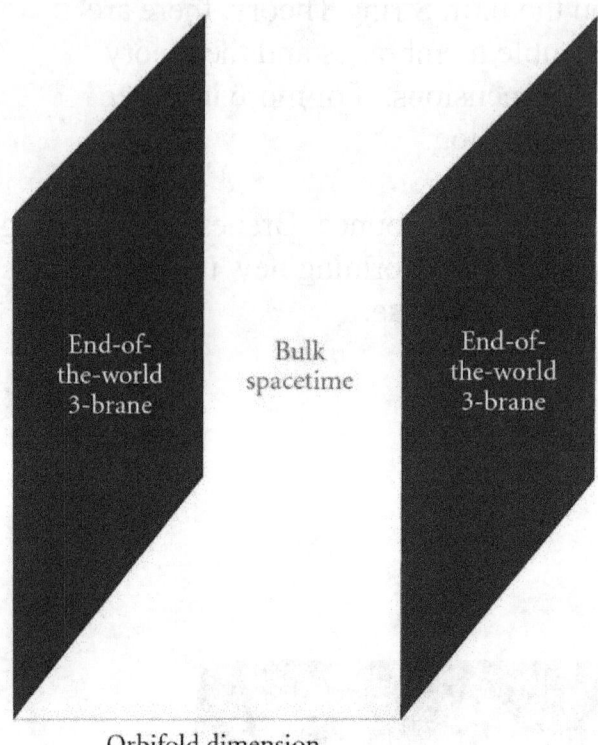

Image: Copyright © 2010 Jean-Luc Lehners. Advances in Astronomy Volume 2010 (2010), Article ID 903907, http://www.hindawi.com/journals/aa/2010/903907/ Creative Common License

6) Ultimate Multiverse Theory: Basically a variety of the above exists. A lot of this is theoretical, but there is a lot of scientific support and mathematics endorsing these theories. (For more info got to http://space.mit.edu/home/tegmark/)

8. What comic books like to delve into the multiverse?

In 1985 and 1986, the writers and designers of DC comics, put out a second series of comics that involved the Big Bang creation of multiple universes which included parallel Earths. Within these Earths were multiple versions of the Flash, Superman, Green Lantern, Wonder Woman and Batman. On each Earth, the personalities, the environments and the relationships were different for each duplicate universe and world and not every character existed. The name of the series was Crisis on Infinite Earths.

Randall A. Golden

One television show that makes use of parallel universes is Fringe, on Fox, They differ politically and technologically. There are duplicates of each person on each parallel world of the earth. The Fringe unit, itself, investigates strange, paranormal events that seem to transcend conventional exploratory methods used by the government. No spoilers. Watch it on iTunes.

Another one of my favorites is a movie called Sliding Doors which is about two parallel universes where the protagonist behaves differently in her relationships. It stars Gwyneth Paltrow. In real physics, parallel worlds would not be connected, so what you would be watching, on the film, is an examination of one universe to another and then going back and forth between both. The same character would react differently on each universe. The protagonists would not be aware of the each other.

Photo: Source: Fashioncraz.com

9. How would a parallel world or universe affect me?

It would not. You might not exist, or you would be a drug dealer on one world or President of the United States on another. Under this theory, the number of universes would be infinite, so anything is possible. Polygamy could be legal and you could have six wives or the concept of marriage may not even exist. It is better that you did not know your other selves.

Now that you are aware of how the universe began and what it is, let us look at the scientists that began observing, mapping, hypothesizing and creating laws about the universe, solar system and our earth.

10. Nicolas Copernicus (1473-1543)

His theory was that the Earth rotated daily on its axis and revolved around the Sun yearly. He managed to document all this with the naked eye because the telescope was not invented yet. This was known as the heliocentric system, and it was critical of the Ptolemaic theory (everything revolved around the Earth, including the sun). Copernicus' work was finally published in 1543 and not accepted until around 1835. The Index was a list of books put out by the Roman Catholic Church that they considered immoral and dangerous. His book was on it.

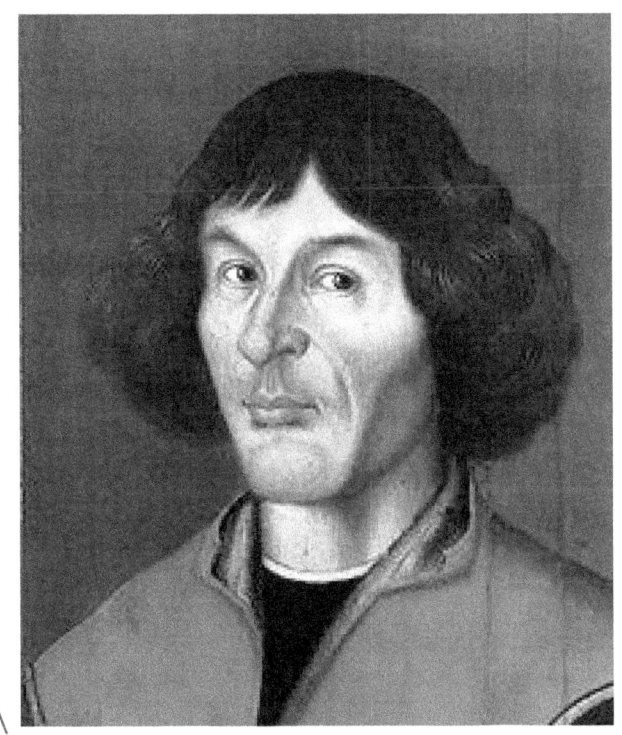

http://www.frombork.art.pl/Ang10.htm Photo Public Domain

11. Galileo Galilee (1564-1642)

Photo: Public Domain
http://www.nmm.ac.uk/mag/pages/mnuExplore/PaintingDetail.cfm?ID=BHC2700

When Galileo had discovered that the telescope had been invented, he went about constructing one himself. He was the first person to use the telescope in the study of astronomy and view craters on the Moon. The moon had previously thought to have had a smooth surface, so this was a big discovery for him. He alternatively discovered four of Jupiter's satellites; saw the Sun spots, Saturn's rings and the phases of Venus. Galileo had a unique and creative way of making telescopes and as a result, there was quite a demand all through Europe for his prolific design, which was sensitive to discovering cosmological entities, in the night-time skies.

He published his findings and found himself in at odds with philosophers and professors. The idea that the earth moved around the Sun was not in accord with bible and in 1614, he was denounced to the Inquisition by Dominican preachers.

Like Copernicus, Galileo's books were found, in the Index of Forbidden Books, in 1616. This was not the kind of publicity he wanted as an astronomer. This was of course the beginnings of the anti-science movement. Galileo was instructed not to teach the Copernican system, but he could discuss it as a hypothetical theory.

In 1624, Galileo received permission from Pope Urban to write a book on systems of the world but only if he came to the conclusion that satisfied the Pope. Galileo tried to be flexible with his conclusion within the book.

He also changed the title, but the Pope called him before the Inquisition to stand trial for heresy.

Galileo was tried in Rome, in February 1633 and on June 16, and was sentenced and found guilty. He was ordered to recant and was sentenced to life imprisonment which was commuted to house arrest. Galileo remained under house arrest for another eight years until his death on January 8, 1642.

In 1998, the Roman Catholic Church admitted it had been wrong about Galileo!

12. Charles Robert Darwin (1809-1882)

Darwin inspired biologist, Jerry Coyne to write his famous blog "Why Evolution is Real."
http://whyevolutionistrue.wordpress.com

At age 22, he went aboard the HMS Beagle as an unpaid naturalist for a voyage around the world, where he went for 5 years with much time spent around South America. He collected many natural environmental specimens and wrote exhaustive notes in a diary and journal. These voyages would change our world perspective in the biological progression of the species. In 1839, Journal of Researches was published and following his controversial book On the Origin of Species by Means of Natural Selection was published in 1859. It basically stated that life evolves by a process of **evolution, powered by natural selection** and random mutation. The theologians were quick to attack him because they could not find God's role in his scientific conclusions, which also contradicted special creation, a concept originating from the bible.

Public Domain

Source
- Charles_Darwin_seated.jpg

Author
- Charles_Darwin_seated.jpg: Henry Maull (1829–1914) and John Fox (1832–1907) (Maull & Fox)
- derivative work:User: Beao

Terms: Natural Selection is the process where organisms have successful genetic characteristics that are appropriate for their survival, in their environment, and are genetically passed on while less suitable traits slowly disappear, in future generations.

Random Mutation: These are DNA mutations that produce characteristics that are not necessarily based on whether they would be useful or not. A mutation is a change in the DNA. The DNA controls the characteristics of the organism such as physiology, appearance, behaviour and protective features. A random mutation may occur from the environment or manifest for no reason. Do not get mixed up with Natural Selection which is not random. The process selects the winners over generations. Evolution is not random.

   Lastly the concept of evolution took place before Darwin; however he found the mechanism of Natural Selection.
http://www.aboutdarwin.com/literature/Pre_Dar.html

13. Why was Isaac Newton's perception of the world, so important?

http://www.newton.cam.ac.uk/art/portrait.html Public Domain

He had a Deterministic Worldview (that all events, physical and mental have a root cause and follow logic and not chance) that was supported by most scientists, at the time, but would change after Einstein and Quantum Physics.

Newton studied at the University of Cambridge, England and was responsible for creating a highly evolved scientific culture for that era. In 1867, he published Philosophiae Naturalis Principia Mathematica a scientific works called the Principia in 1867. Newton`s crowning achievements were his three laws of motion and his universal law of gravity. The universal law of gravitation stated the universe moves like clockwise in various ways. It is a myth that an apple descended from the heavens and collided with Newton`s head inspiring him to have an epiphany regarding gravity. Actually Newton tended to observe the falling of apples and connect this observation to the earth's movement and the moons movement around the earth. He discovered the tool of fluxions, which we now classify as calculus, and he found there was a force of attraction between all objects with mass in the universe. There are calculations to figure out the force between the two. You multiply the mass of the two objects. Distance has an effect on gravitational force, as well. As you move away from an object, the force becomes less. You always experience gravity from the centre of the earth, and there is always gravity in space. Newton`s Laws of Motion included Inertia, Motion and Reciprocal Action. He created the reflector telescope which inspired astronomers for hundreds of years

Newton`s Telescope Photo: Author: **Szőcs Tamás**
Tamasflex

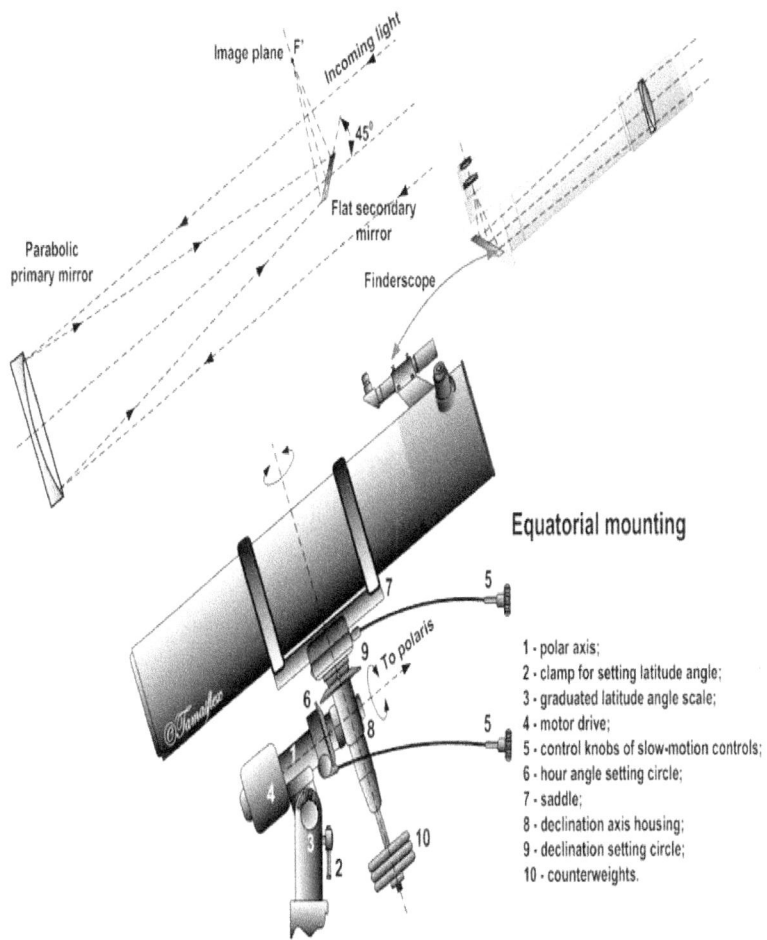

Will Kalif, from Telescope Nerd, describes the telescope as a reflector telescope. "A reflector telescope is one that uses a mirror rather than lenses to bend light and magnify images. Reflector telescopes, because they are easier to make and can be made in sizes much larger than refractors, are an invention that changed astronomy and our understanding of the universe. The largest refractor telescope in the world is forty inches in diameter and reflector telescopes dwarf this in comparison. There are currently several reflector type scopes that are over four hundred inches in diameter."

http://www.telescopenerd.com/astronomers/sir-isaac-newton.htm

14. Let's review. What formula created this and who was one of the world's most brilliant scientists that created the formula?

U.S Federal Government Public Domain

Answer: $E=MC^2$. In 1905, Einstein, in his Relativity Doctrine, professed the equalization of mass and energy. He found that as an object converged on the speed of light, the mass increased and it was impossible to get to the level of the speed of light. Mass and energy were codependent. So that is how the equation from left to right works. If we go right to left, we convert mass to energy creating atomic fission. Astronomers see fission on our sun every day, and we are using the idea to produce nuclear fusion for the future. Radioactivity is the perfect example of the conversion of mass to energy. So $E=mc^2$ totally surrounds us in the form our smoke detectors in our homes and the use of radio-isotopes used in medical scanning devices and not to mention nuclear power. You should know that ionized radiation is the radioactive radiation that causes cancer. CAT, MIR scans, the sun and various x-ray including dental. Non ionized radiation does not trigger cancer because it is too weak to violate the DNA bond and cell structure. That is why you should not worry about cell phones or WIFI. However, the sun`s rays are non ionized and may cause cancer. Before you jump on the bandwagon of illiteracy, do some studying.

15. What is Relativity?

In 1905, the world changed with the publication of Einstein's Theory of Relativity.

Special Relativity stated the physical laws like space and time are not independent concepts but are a connected form; a continuum called the space-time continuum. Waves and particles cannot move freely but they must move forward in time, in the Space Time Continuum, while not going faster than the speed of light. We also have mass-energy equivalence which $E=mc_2$ which means that mass or matter can be transmuted in pure energy and the formula can be applied to nuclear energy. Phenomena called time dilation affects clocks; which means that moving clocks measure time slower than stationary objects. In relation to the observer, objects are measured to be shorter in the direction in which they are travelling, and this is called length contraction. It was actually Max Plank who labelled Einstein's theory of Special Relativity.

The effect of matter or mass on space-time is to tilt the direction of space-time towards the matter itself. The objects tend to move towards the masses and this kind of tilting of space-time is called gravity. What we find is that the curving or bending of space-time is producing gravity. The amount of tilt effect (or degree of curvature) is proportional to the mass and the closer you get to the mass; the tilt effect becomes greater and the more intense the gravity. Think about throwing a heavy metal basketball (Earth) on a trampoline and that is the effect.

When an object is dense enough and massive enough, the tilt will be so huge that you will reach a point where all possible paths forward in the space-time continuum will go to the centre of the mass, and the object becomes a black hole with the centre being the event horizon. Stephen Hawking, however,

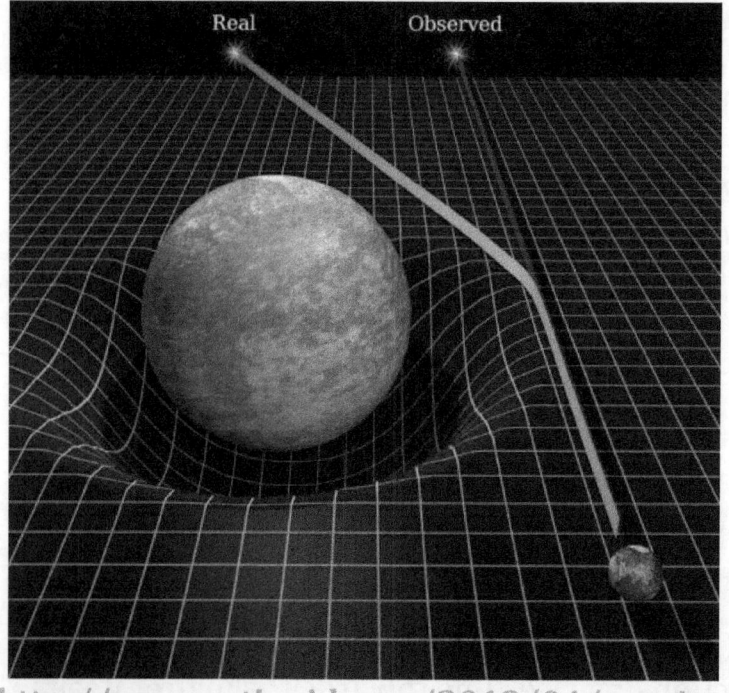

http://www.artinaid.com/2013/04/gravity/

is now critical of the existence of the event horizon.

In General Relativity, the Gravitational Time Dilatation effect, is where time gets slower, as the gravitation force becomes stronger and time increases as you get away from gravity. Momentum or speed is different, as there is increasing momentum and the slower the progression of time and the reverse. Curvatures in the Space-time Continuum can bend light in the presence of a massive object such as a star or a black hole. The universe is expanding due to an unknown force which, Einstein considered the cosmological constant, but he abandoned that idea. He decided there was no expansion. This was a mistake because now we think that the cosmological constant may be dark energy; which may explain the reason the universe continues to expand after inflation stopped.

16. What did Einstein really say about God and religion?

Einstein's last words were to his Jewish Philosopher Professor Gutkind went like this: "The word god is for me nothing more than the expression and product of human weaknesses, the Bible a collection of honorable, but still primitive legends which are nevertheless pretty childish. No interpretation no matter how subtle can (for me) change this. For me, the Jewish religion like all others is an incarnation of the most childish superstitions. And the Jewish people to whom I gladly belong and with whose mentality I have a deep affinity have no different quality for me than all other people. As far as my experience goes, they are no better than other human groups, although they are protected from the worst cancers by a lack of power. Otherwise, I cannot see anything chosen about them."

http://www.guardian.co.uk/science/2008/may/12/peopleinscience.religion

17. Who was brilliant scientist of this century, was gay, victimized and possibly committed suicide?

Alan Turing was born in June 23, 1912. He was an atheist and gay. He suffered for embodying those beliefs, in the intolerant times of the 1950`s, in England. He was arrested for indecent acts twelve times and subject to forced hormone treatments and barred from working. The stresses eventually lead to a possible suicide. We do not know for sure because the evidence surrounding his death is inconclusive. These terrible atrocities were directed at one of the most brilliant scientists, to live on this planet, Mr. Alan Turing. Alan Turing was a British mathematician, cryptographer, and computer scientist who was recognized as the founder of computer science and artificial intelligence. In 1936, he developed the concept of the Turing Machine. He developed a method for analyzing machine intelligence, the respected Turing Test. During World War II, he was credited the success for cracking the German Enigma cipher, which was a major contributing factor to winning the war.

http://commons.wikimedia.org/wiki/File:Alan_Turing_Memorial_Closer.jpg

18. What was the Turing Test?

The Psych. Dept of the University of Toronto describes the test as "there was a human, of either gender, or a computer at the other terminal. Now the judge's job is to decide which of the contestants is human, and which the machine. Turing proposed that if, under these conditions, a judge were less than 50% accurate, that is, if a judge is as likely to pick either human or computer, then the computer must be a passable simulation of a human being and hence, intelligent. The game has been recently modified so that there is only one contestant, and the judge's job is not to choose between two contestants, but simply to decide whether the single contestant is human or machine. This is a test for viable artificial intelligence." [6] Here is his home page.
http://www.turing.org.uk/turing/
http://www.fil.ion.ucl.ac.uk/~asaygin/tt/ttest.html

---

[6] http://www.psych.utoronto.ca/users/reingold/courses/ai/turing.html

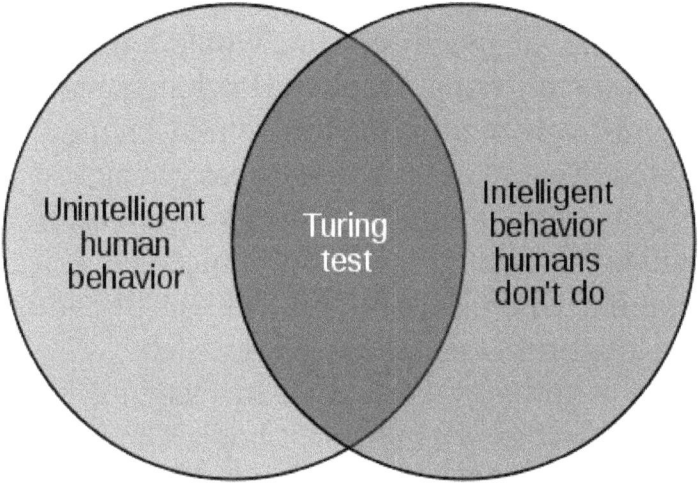

Source en:Image:Weakness of Turing test 1.jpg
Author en:User:CharlesGillingham, User:Stannered

19. Why are Stephen Hawking's accomplishments so amazing?

He was born on January 8, 1942 in London, England and went to Cambridge, in his late teens, to study the universe, and its laws. With Roger Penrose, he demonstrated that Einstein's General Theory of Relativity implied that space and time would have a beginning with the big bang. Stephen Hawking's main goal was to find a way of unifying Einstein's General Theory of Relativity; the other three forces (weak, strong, electromagnetic) with Quantum Mechanics. He is best known for "Hawking radiation" which is the discovery that black holes are not really black but emits energy due to quantum virtual particles which constantly pop in and out of existence in a vacuum (which we have discussed). In his last book, The Grand Design, he talked about "spontaneous creation" of the universe without necessity of God, which caused controversy but was supported by most cosmologists, who are atheists. There was also controversy whether information could survive the singularity of the black hole. He made a bet stating it could not and lost. We also learn about how the information is stored through the Holographic Principle (data can be stored on a flat two dimensional object called the event horizon and part of a black hole and then retrieved in 3D. The Holographic Principle has been upgraded to include, a theory that the universe is a whole hologram.

http://www.nature.com/news/simulations-back-up-theory-that universe-is-a-hologram-1.14328   There are scientists that disagree.

http://www.universetoday.com/107172/why-our-universe-is-not-a-hologram/#.UquL-mywajM.facebook

Check out Leonard Susskind's book called "The Black Hole War: My Battle with Stephen Hawking to Make the World Safe for Quantum Mechanics". This book tells the whole story about Hawking`s bet about information and black holes.

Eventually, due to Hawking radiation, all the mass-energy inside a black hole will evaporate. He is a member of the National Association of Scientists and the Royal Society.

NASA PUBLIC DOMAIN

Hawking has recently filed a paper that Black Holes are structured differently compared to what he first had envisioned. He believes there are no event horizons in black holes. http://io9.com/stephen-hawking-says-there-are-no-black-holes-1508408816

These are four of the world's top scientist that have formulated and constructed the scientific nature of the universe and our world for us. A literate nerd should also know some scientific terms related to the universe.

20. What is the force that is responsible the wear and tear on your joints?

Gravity! Gravity is going to make this rocker land on top a group of rockers at this rock concert because gravity is an attractive force between objects of mass.

Source   Dead Fish

Author   [][tatu] from Porto Alegre, Brasil

The power of this force is based on the mass or weight of the object and is inversely proportional to the squared distance between the objects. This mass curves space-time. So the earth attracts us to the center of it, and we never fall off it. The moon orbits the earth and never crashes into it because of gravity. It is the speed of the moon combined with the strength of the gravitational pull that prevents it from crashing into us. Gravity keeps us spinning around the sun. Thanks to Einstein we know that the bending or the warping and curvature of the time and space continuum that produces this.

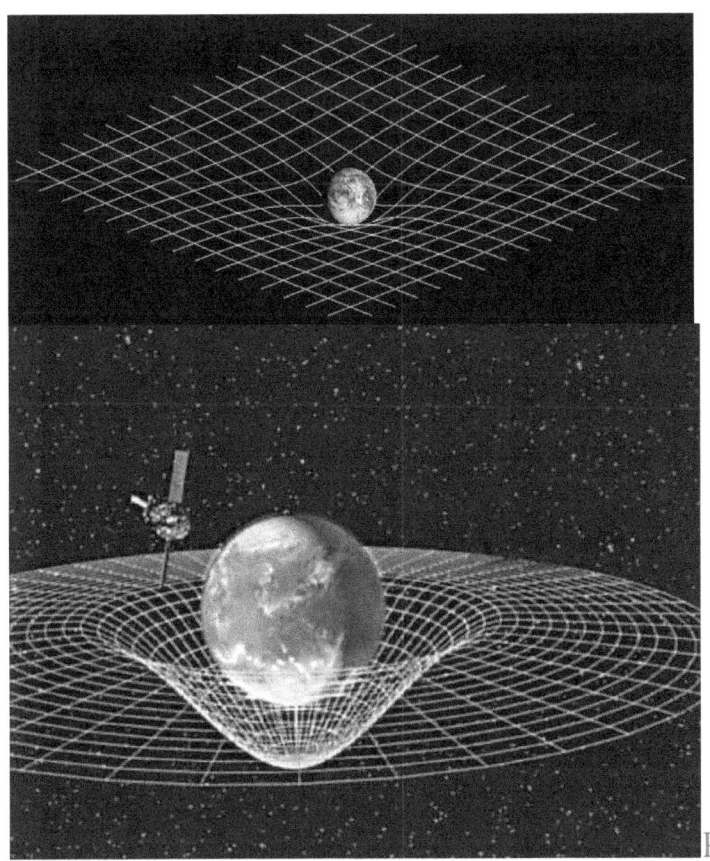

	Public Domain
Source	http://www.nasa.gov/mission_pages/gpb/gpb_012.html
Author	NASA

21. What is the matter with Matter?

In the universe, matter is black holes, dark matter, planets and stars etc. Matter is composed of four states everywhere in the universe. These states are solid, liquid, plasma and gas. A solid is anything that embodies a particular size and shape. An asteroid or planet would be a solid. The only way they can change their shape is by force. A liquid is anything that has size or volume, but does not have a shape. Liquids must be contained in a cup, bottle, or receptacle to have a shape. Gasses have no particular shape of form. They provide the volume of whatever holds them.

In March, 2013, scientists may have discovered salts leaking to the surface that may indicate an ocean of water under the surface. In fact, this water, at Europa, has provoked NASA to want to travel there and explore for possible life due to the trace elements of $H2O$ in the atmosphere. "The icy moon may look tiny next to our own planet, but it's got 2- TO 3-TIMES AS MUCH H_2O as we have here on earth." http://io9.com/this-image-is-why-everyones-so-excited-about-a-nasa-mis-1539388419?utm_campaign=socialflow_io9_facebook&utm_source=io9_facebook&utm_medium=socialflow

Here is more on Gas. It can expand and contract when influenced by changes in temperatures or pressure. There are many gasses in the universe such as Nitrogen, Oxygen, Argon, Carbon Dioxide, Neon, Helium, Methane, Krypton and Nitrous Oxide, which are all found in the atmospheres of many planets. All matter has positive energy. This point will be important when we talk the theory of spontaneous creation from zero energy vacuums. A majority of the matter in space is plasma. The stars that illuminate our skies have plasma. Plasma is a high energy electrically charged mixture of ions and electrons. We can find plasma in locations like lightning, fires, fluorescent light and auroras.

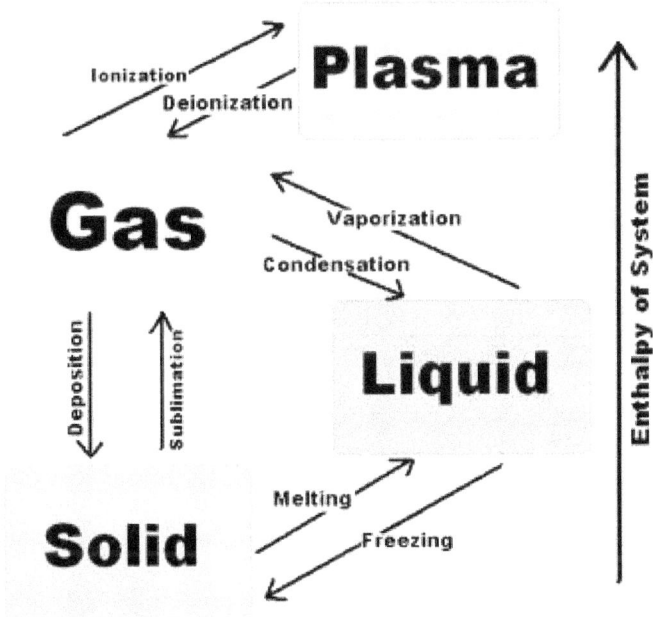

http://science-capt.wikispaces.com/Energy+Transformations

States of Matter — Glenn Research Center

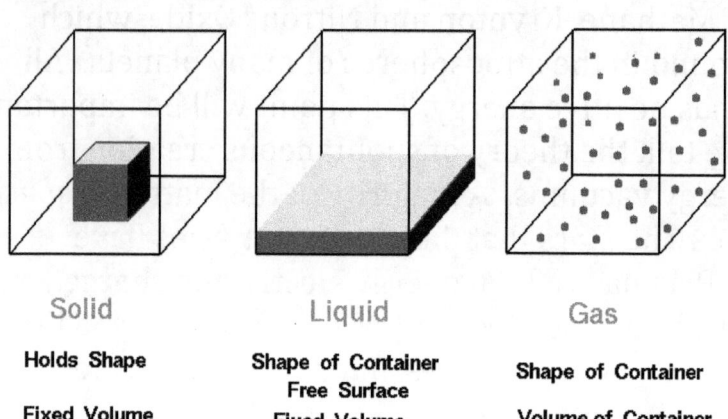

Solid	Liquid	Gas
Holds Shape	Shape of Container / Free Surface	Shape of Container
Fixed Volume	Fixed Volume	Volume of Container

NASA
http://exploration.grc.nasa.gov/education/rocket/state.html

22. What are Dark Matter and Dark Energy and their relation to the Universe?

　　Dark Matter is a type of matter that appears to cling and permeate galaxy clusters. Some evidence indicates that it may be made of WIMPS.

　　NASA describes WIMPS as "subatomic particles which are not made up of ordinary matter. They are "weakly interacting" because they can pass through ordinary matter. WIMPs act as their own antimatter particles. When two WIMPs interact; they annihilate each other. The interaction causes a release of secondary particles that emit gamma rays. Using GLAST, scientists hope to find these high-energy signatures of dark matter in our galaxy. If they succeed, this discovery will help solve one of astronomy grandest mysteries."
http://www.nasa.gov/mission_pages/GLAST/science/dark_matter_prt.htm

　　Dark Matter is approximately 83 percent of the matter of the universe and 22 percent of the universe. Other matter includes luminous matter which can be seen and is not transparent. Recently, they have not found much around the stars, so they think it is going to be difficult to find particles on Earth.
http://www.sciencedaily.com/releases/2012/04/120418111923.htm. Dark matter was thought to form a type of glue for the universe and a haze around the Milky Way which certainly rotates much faster than the visible matter but if there is none around the Milky

Way, new questions open up. As of July 4, 2012, with the Higgs boson being discovered, it is quite probable that the dark matter is WIMPS and this may link the discovery to supersymettry (For every particle, which are called Fermions and Bosons, in the Standard Model, we may have undiscovered cousin particles.) Lately scientists have found negative evidence against supersymettry which includes the strange shape of electrons and the non discovery of some particles that should have shown up at the Hadron Collider, at low energies, have not appeared.   Great article at http://scienceblogs.com/startswithabang/2013/05/15/the-rise-and-fall-of-supersymmetry/ .

   These cousin particles are a duplicate to the known particles in every way except that they have a different spin. Also if suppersymettry is weak as a theory, then we may have a problem with String Theory. http://www.scientificamerican.com/article/is-supersymmetry-dead/

   Dark Energy is related to what Einstein believed; that there was this repulsive force that counteracted gravity, and he called that the cosmological constant. He believed that was responsible for causing the universe to expand but then he discarded the idea. He was wrong to discard that concept because that was most likely what we refer to as the cosmological constant and may be dark energy. This would account for the expansion, long after the big bang and inflation. It turns out that roughly 71 percent of the Universe is dark energy depending on your research.

Latest NASA image:

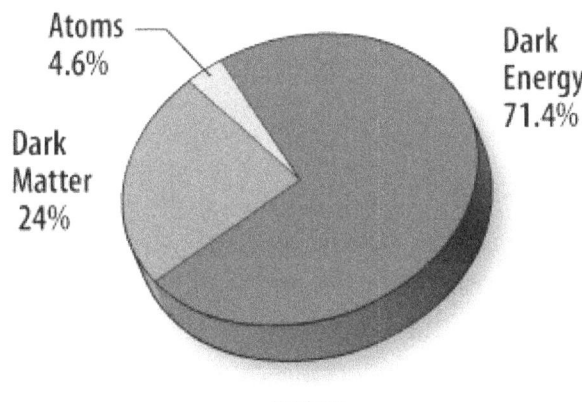

Credit: NASA/WMAP Science Team

23. What is a vacuum?

Vacuum: It is the absence of matter and time with the lowest possible energy. It can occur on a subatomic level in the production of virtual particles and it can be part of the spontaneous creation of universes, in the space, where zero point energy vacuum states create quantum fluctuations. On a subatomic level, (the microscopic world), the quantum vacuum always has a minimum of electromagnetic waves embodied in it, allowing the spontaneous popping in and out of the virtual particles, made up of antiparticles and standard particles. All this was discovered by Dirac as part of Dirac`s Soup and Quantum Field Theory. Another effect connected with this vacuum is called Casimir's effect. The Dutch Physicist, Hendrik Casimir, in 1948, found that if you have two uncharged metallic plates in a vacuum or electromagnetic field, the fluctuating quantum vacuum electromagnetic waves or fluctuations will set up an attractive force between the plates. Virtual particles will emerge. If we total the energy of all the virtual particles in the vacuum to the sum, it will work out to be zero point energy. This energy total would be valid without the plates. This is a legitimate way of testing for the existence of the vacuum. The Zero energy environments allow for the virtual particles to hang around without annihilating themselves for longer periods of time. This is important when you take this concept into creating universes. Yes, I know this is repetitious but necessary and I can see the brightness in your eyes as you continue to take in the concepts.

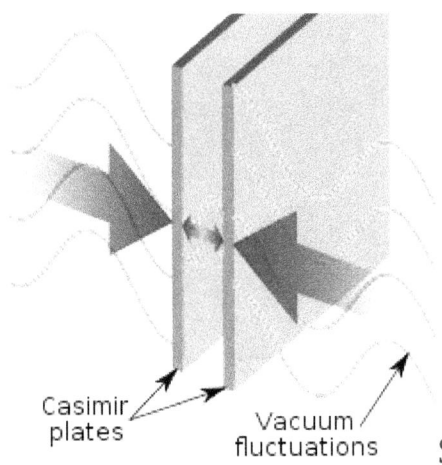

Source: Emok Photo: http://commons.wikimedia.org/wiki/File:Casimir_plates.svg?uselang=es

The link, http://www.scientificamerican.com/article.cf=what-is-the-casimir-effect discusses the Casimir effect, if you want an excellent and more advanced look.

    A vacuum can occur in outer space, as an area where there is no almost zero energy, no time matter or gravity. This is an area where there is nothing and something comes from the vacuum while the negative energy of gravity has balanced out the positive energy of the surrounding matter creating almost zero energy. Many scientists theorize the big bang starts from here, and a quantum vacuum fluctuation forms the universe. We will examine that later.

24. What is another word for the Three Stooges? Entropy

Wikomedia Commons
http://commons.wikimedia.org/wiki/File:Disorder_jamison.jpg Public Domains

Entropy is described as the development from order to disorder, from useable energy to unusable energy, from a solid piece of wood to a pile of ashes.

25. What does Entropy have to do with Time?

Much of what I learned about Entropy and time came from Sean Carroll's lectures and books on Time. I will take aim and simplify it for you, but I highly recommend his book called "From Eternity to Here"

The second law of thermodynamics states that the disorder in the universe always increases and that leads us back to entropy which is the energy from order to disorder. The theory is that time actually does not go forwards or backwards and the only reason we perceive it, as doing so is because of a scientific concept called Entropy. The normal progression in the universe is from low entropy to higher entropy. However, nothing in the laws of entropy says it cannot be the reversed when a universe comes into creation. In the Marvel Universe, there is a character called Entropy who possesses unique powers of creating disorder. He was a representation of Eternity born at the beginning of time, whose ambition was to unravel, the cycle of creation and destruction will forever continue.

http://marvel.com/universe/Entropy

In our universe, entropy is as described by the famous physicist Sean Carroll, "Entropy is a measure of the DISORDER of a system. A nice organized system, like an unbroken egg or a neatly-arranged pile of papers, has low entropy; a disorganized system, like a broken egg or a scattered mess of papers, has high entropy. Left to its own devices, entropy goes up as time passes. The more things we can do to atoms, the higher the entropy." [7] The universe, which was an open system, started off with low entropy and is now high. The universe has expanded with the help of inflation and is possibly maintained with the cosmological constant dark energy. We make the time direction or the "Arrow of Time"[6] by doing high entropy activities like pouring wine into a glass, cutting trees and washing our car because primarily the fundamental laws of physics see the past and the future as exactly the same, but you and I perceive an actual direction of time brought on by entropy. From this account, you cannot go backwards and change anything. Entropy provides us with our management and story in life, and it is the disorder behind the creation of our timeline on Facebook. In the multiverse, Carroll speculates that a universe created with high entropy that would evolve to low entropy. "It will be just like the observed history of our universe — just played backward. A collection of long-wavelength photons will gradually come together and radiation will focus on certain locations in space to create white holes. Those white holes will spit out gas and dust that will form into stars and planets. Radiation will focus on the stars, which will break down heavy elements into lighter ones and eventually all the matter will disperse as it contracts and smoothes out to

---

[7] http://preposterousuniverse.com/eternitytohere/faq.html

create a giant Big Crunch. Along the way, people will not die, grow younger, and be un-born; omelets will convert into eggs; artists will painstakingly remove paint from their canvases onto brushes" [8]

This resembles the Bizzaro World for those who follow Superman. In the Bizarro world of ""htrae, (EARTH), the civilization is ruled by the Bizarro Code, which involved the cultural opposite of the value system of the regular earth. There were Bizzaro copies of Superman and all the superheroes with reverse vulnerabilities and strengths. The twisted planet was actually shaped like a cube, rather than an earthly bubble-like sphere, differing itself on a cosmological basis. Ugly facial features and ridiculous psychological behavior were reinforced, while godly deeds and perfection were discouraged by peers and authority. Also think of the movie and book, *The Curious Case of Benjamin Button* with Brad Pitt based on the book written by F. Scott Fitzgerald. It is the story of a man who is born in his eighties and has developed the tastes of an older person, but who now begins to physiologically evolve to a younger state, all the way to a baby. We see the regression and progression emotionally and intellectually. This is a touch of high to low entropy in our universe on a fantasy level.

---

[8] http://blogs.discovermagazine.com/cosmicvariance/2011/08/03/a-universe-out-of-chaos/#.UoUUxvtFPRg

http://illinoisreview.typepad.com/illinoisreview/2012/01/whatever.html

http://olympusenglish3.wikispaces.com/The+Curious+Case+of+Benjamin+Button

The world of Quantum Physics or the subatomic world is a random unseen world filled with laws and theories that we have yet to discover.

26. What is a simple explanation of Quantum Physics?

From 1900-1913, Max Plank and Albert Einstein proposed that photons or light were waves and particles and Neils Bohr constructed a model of the atom based on quantum ideas. It resembled our solar system.

The best definition of Quantum Physics came from a New York Times Article by the famous Physicist Brian Greene, in 2005, called One Hundred Years of Uncertainty. Quantum Physics deals with the random, nondeterministic behaviour of particles on a subatomic level.

Greene wrote, "An example: Imagine you shoot an electron from here and a few seconds later it was detected by your equipment over there. What path did the electron follow during the passage from you to the detector? What the answer according to quantum mechanics? There is no answer. The very idea that an electron or a photon, or any other particle, travels along a single, definite trajectory from here to there, is a quaint version of reality that quantum mechanics is declared outmoded. Instead, the proponents of quantum theory claimed, reality consists of a haze of all possibilities -—all trajectories -—mutually commingling and simultaneously unfolding." And why don`t we see this? According to the quantum doctrine, when we make a measurement or perform an observation, we force the myriad possibilities to ante up, snap out of the haze and settle on a single result. But between observations -—when we are not looking

-reality consists entirely of jostling possibilities." [9] That is particularly illustrated in the double slit experiment.

If you find this is too difficult

Try to remember that all particles are random and not able to be determined on a subatomic level and that they embody wave and particle duality, until measured or observed. Just that one sentence will impress anyone.

27. Here is a picture of the atom and its parts.

---

[9] http://www.nytimes.com/2005/04/08/opinion/08greene.html

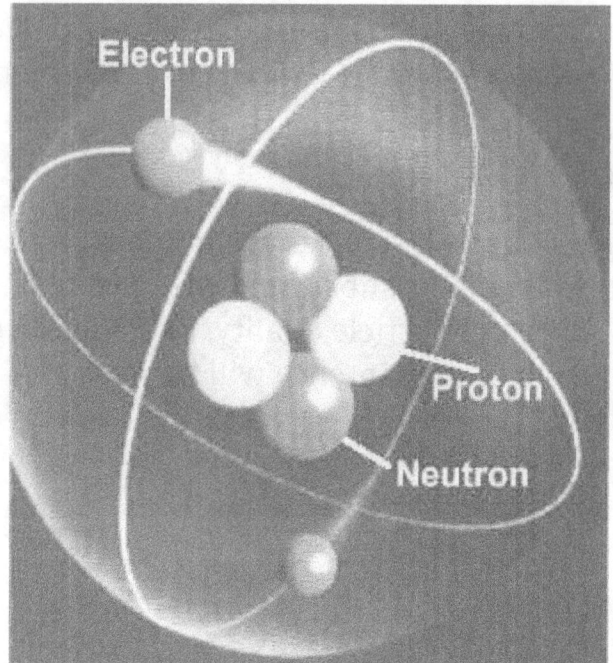

http://pryken.wikispaces.com/Max+and+Vlad%27s+atoms

ATOM: It is the smallest part of an element consisting of a nucleus consisting of combinations of neutrons and
protons and one or more electrons attached the nucleus by electrical attraction.
PHOTON: Particles of light that carry the electromagnetic force.
ELECTRON: Negative electrical charge. Flow of electrons is electricity.
PROTON: Positive electrical charge and is in the nucleus of the atom.

We should also mention that wave-particle duality holds that light and matter exhibit properties of both waves and particles.

On May 20, 2014 an amazing bit of research was published. Scientists are hypothesizing that colliding photons and could produce electrons and positrons. Basically they are writing about converting light to matter. This is a dream that has been in the mind of scientists for the last eighty years since the creation of the atomic bomb.

http://www.theguardian.com/science/2014/may/18/matter-light-photons-electrons-positrons

Al Pratt and Ray Palmer became The Atom in DC`s ATOM series comics. He was able to shrink in size on a subatomic level to challenge evil and fight for justice against the nefarious evil criminals.

 The Dooms from Beyond!

http://www.comics.org/issue/16544/Atom / comic story /DC Credits Pencils: Gil Kane

28. What did the double slit experiment demonstrate and teach us about quantum mechanics or physics?

A Quantum Event: Double Slit Experiments: Wave Particle Duality Simplified: In 1802, Thomas Young, a scientist showed that when you passed light or photons (chunks of quantified light; a name given to the light after 1905 by Einstein, when he discovered the photoelectric effect and declared light as particles) between two slits, the light behaved like waves and interfered with each other creating bands of light and dark. Later these would be found as probability waves where the photons interfered with each other, went through both slits at the same time and had a wavelength which would later on be interpreted by Schrödinger as a wave function, indicating all the possibilities where there could be photons. Of course, once measured or observed, the wave function would collapse and light was back to chunks or photons.

The Photoelectric Effect won Einstein the Nobel Prize, and it described the idea of how a light source, when shined on a metal surface, could force the electrons to be ejected from that surface. Electrons emitted were then called photoelectrons and proposed that the light was to be called photons and were particles. What they had discovered was wave/ particle duality. Light could be waves and particles. Light waves going through one slit did not interfere with each other, but they were diffracted. When they went through two slits, they were wave-like; demonstrating the probability or probable positions that the photon particles could be on the wave function and, as a result, they appeared like interference patterns. However, when photons were observed or measured going through two slits, they did not interfere with each other and behaved like particles, not waves, going through two slits and their possible locations on probability waves ceased with the collapse of the wave function.

In 1923, D. Broglie used the same experiment with particles called electrons. With one slit, the electrons acted like particles making one bar on the wall, but when there were two equal slit choices (a quantum event), the particles again behaved like wave –like probabilities on the wave function, creating an interference pattern of probabilities. Observation or measurement created a collapse in the wave function. That is what Erwin Schrödinger called the quantum wave function. John Gribbin, in his book, In Search of Schrödinger's Cat, stated that each "ghost" electron corresponds to "a wave or packet of waves that describe one real electron." [10] There are many of these ghost electrons until during this interference cycle. The ghost particles would continue to interfere with each other until the collapse of the wave function by measurement or observation and thus producing and going back to "the one real particle." [11] Stop looking at the process and the real electron again spits into its many ghost electron particles.

---

[10] John R. Gribbin, In Search of Schodinger`s Cat(New York, N.Y:Bantam, 1984)

[11] John R. Gribbin, In Search of Schodinger`s Cat(New York, N.Y:Bantam, 1984)

Quantum Field Theory and quantum events would again be applied in the inflationary theories that demonstrated creating universes from nothing. Quantum physics is deterministic in terms of law but random in behavior. Heisenberg noticed that if you measured or tried to change the momentum (velocity) or position of one of these electrons or photons, it effected the other one so you could not simultaneously do both which became Heisenberg`s Uncertainty Theory. E.g. If I try to change the momentum of a photon, the position is unavailable to me and the reverse.  If it was a ball, the speed or momentum of the ball would be available but I could only discover a probability of the balls position. Again, this is called Heisenberg`s Uncertainty Theory. Again, an excellent video which describes the theory in one minute is this one. http://www.youtube.com/watch?v=7vc-Uvp3vwg

http://ibphysicsstuff.wikidot.com/two-source-interference

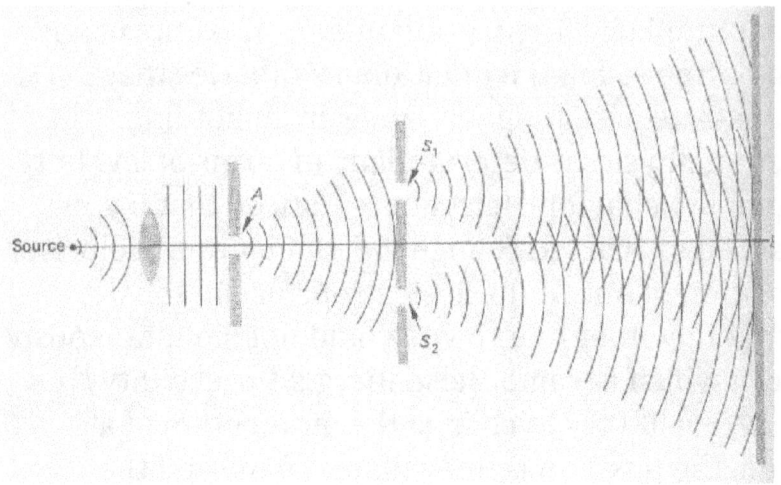

http://ibphysicsstuff.wikidot.com/two-source-interference

29. Schrödinger's Cat 1: How can a cat be alive and dead at the same time?

How can a cat be alive and dead at the same time? Let`s review this again. The answer is superposition as mentioned in the above video, if you went to the internet.

Schrödinger's Cat:
   Schrödinger always felt that quantum physics was so weird that it could never happen from a macroscopic perspective in real life, so he conjured up this imaginary scenario. A cat is put in a box with a radioactive nucleus, which has a fifty –fifty chance of decaying in hour, due to a small bottle of deadly poison.

If the nucleus decays, it shatters the glass and bottle while poisoning the cat and the cat is dead. If it does not decay, the cat is alive. A fifty-fifty chance illustrates a quantum event on a subatomic level and applies this to the cat's dilemma. In traditional quantum mechanics, the cat is now attached to the wave function which exhibits states of alive and death and in a state of superposition. There are two ways to solve this. Open the box and the wave function collapses which allows us to see the cat either dead or alive. (This is part of the Copenhagen Interpretation). The Many Worlds Theory by the late Hugh Everett would say there is no wave function and that when you opened the box and found a dead cat, you have created a parallel world with one world having a live cat and the other a dead cat.

What happens when you have a quantum pigeon and cat?

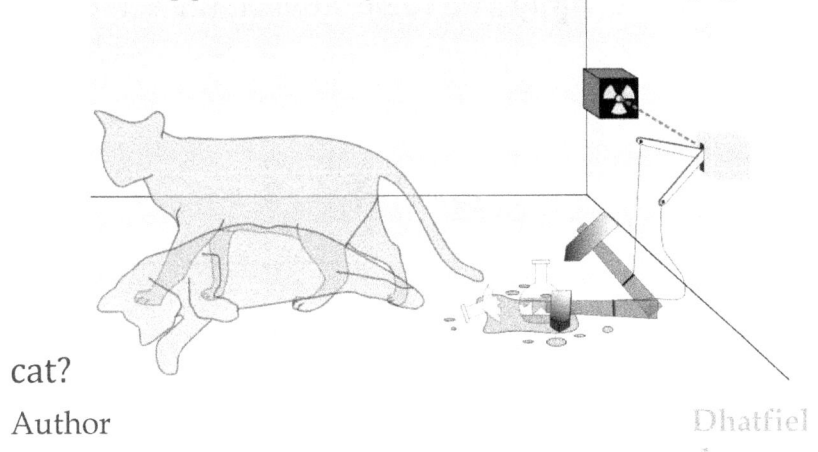

Author Dhatfield

Permission CC by SA 3.0

http://en.wikipedia.org/wiki/File:Schrodingers_cat.svg
http://phys.org/news/2014-08-physicists-quantum-animal-pigeon.html

    Another famous cat was Felix the Cat, who was created 1919, in Feline Follies and was nearly as famous as Mickey Mouse. The Prince of Wales picked Felix the Cat as his Polo Team mascot and a picture of Felix accompanied Charles Lindbergh in his historic solo-flight across the Atlantic.

    In 1923, a Felix the Cat comic strip was created and ran in newspapers until 1943. It was very popular during WWII. The audience was amazed at the first animated cat that appeared to be alive on the screen.

P137137ublic Domain Otto Messmer, converted to vector by Tom Edwards
http://en.wikipedia.org/wiki/File:Felix_the_cat.svg

30. Who is Feynman and why was he so respected by his peers?

   He was considered one of the most famous Quantum physicists. Richard Phillips Feynman was born on 11 May 1918, in Manhattan, USA. Feynman integrated Quantum Field Theory and Electromagnetic Theory. He called it QED or Quantum Electrodynamics Theory. It is the story of how charged particles like muons, electrons and photons interact in an electromagnetic field. It is quantum field theory. Two electrons communicate their electromagnetic information by emitting and absorbing a photon. A photon that acts in this manner is called a virtual photon. He devised specific diagrams relating to how particles interacted. E.g. Electron would emit a photon and the photon would be absorbed by another electron. This was outlined in one of many Feynman Diagrams. Two particles exchange information using a photon messenger. This is called QED or Quantum Electrodynamics.

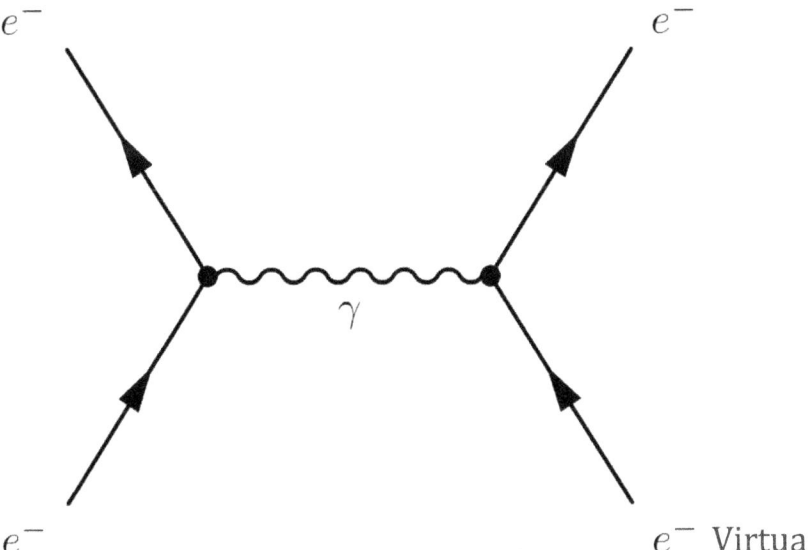

Virtual proton emitted and absorbed by electron. He created diagrams like this one. Feynman made many diagrams like these to describe his theories.
http://upload.wikimedia.org/wikipedia/commons/archive/7/78/20080308114514!Feynmandiagramm.png User: Papa November

Again, he was responsible for expanding on Quantum Field Theory, with Paul Dirac, which included QED Quantum Electrodynamics. Virtual particles, like photons, electrons, positrons or anti-protons, pop in and out of existence for a brief period, thanks to the Uncertainty Principal. The photon is its own antiparticle. As previously noted, another name for virtual particles is quantum vacuum fluctuations which are particle or antiparticle pairs. They carry a large fluctuation of energy. He was involved in the Manhattan Project which instrumental in the creation of the atomic bomb; something he regretted during his lifetime. He produced a series of lectures that eventually became the Feynman Lectures on Physics, and that is one reason that Feynman is still regarded as one of the greatest teachers of physics. In 1982, Richard Feynman thought up the idea of a "quantum computer" which is a computer that uses the concepts of quantum mechanics to increase storage and function beyond the regular conventions of the computer. He developed the famous Feynman Diagrams based on QED and worked with NASA. He was a skeptic and a quiet atheist. He lost his belief in God as a result of the Holocaust. Feynman felt that critical skills and empiricism were important aspects of a person's worldview. His main accomplishment was quantifying electromagnetism. .

Blog Viktor Gómez ""alentinos""by Viktor Gómez blog is licensed under a Creative Commons Reconocimiento 2.5 España License. Based on a work at http:viktorgomez.blogspot.com.

30. Why did Walter name himself Heisenberg in Breaking Bad?

Walter was a science teacher, and he idolized the famous scientist Werner Heisenberg. So he became the meth character called Heisenberg, as homage to Werner, in Breaking Bad, when he was contemplating how much meth he needs to make in his white, shabby, little van. Werner was famous for Heisenberg's Law of Uncertainty. Photons or electrons are hard to find when you are looking for how fast they are moving or where they are. Here is the fancy language. You cannot measure a particles' momentum or position at the same time, or if you know one or the other, you cannot measure the converse. I know the position; I will never know the exact momentum. Heisenberg's Theory of Uncertainty, also prevents virtual particles from visiting too long, in the subatomic world but when we are talking about the universe or universes, zero energy negates the principal, allowing universes to pop in and stay. Position and Momentum cannot be known simultaneously. This Heisenberg Principle video is quite excellent.

http://www.youtube.com/watch?v=a8FTr2qMutA
.

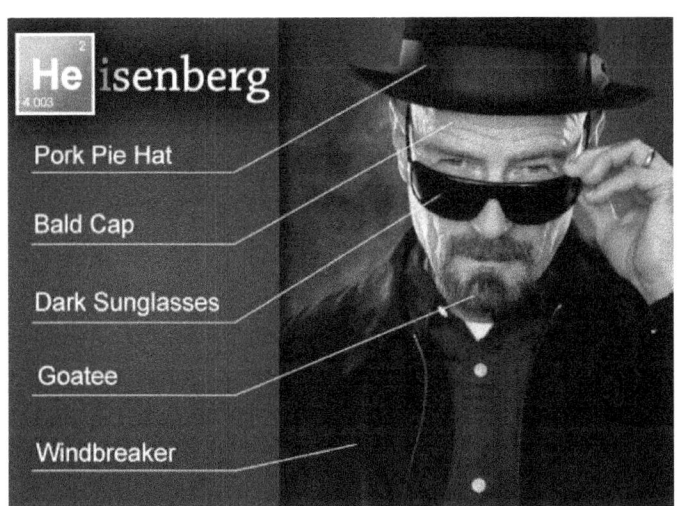

Somos Creative Commons

La Parabólica by La Parabólica is licensed under a Creative Commons Reconocimiento-SinObraDerivada 3.0 Unported License

31. There are no ghosts, spirits, poltergeists, x-ray visions or telekinesis that you see on those crazy reality shows.

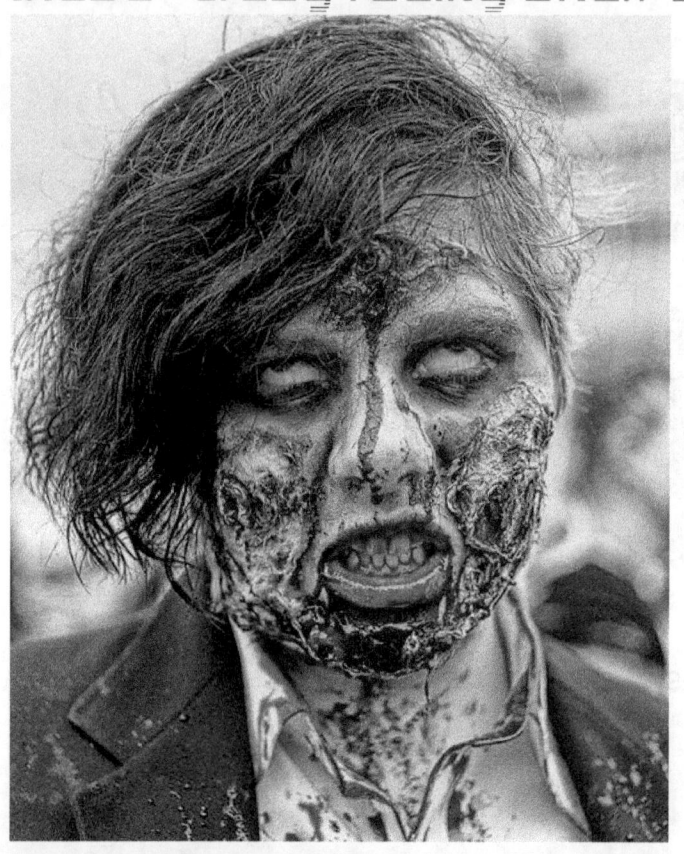

Source originally posted to Flickr as Happy Halloween !-1

Author Bob Jagendorf

I will tell you why. The concept violates the laws of Quantum physics. The investigators are looking for weird energies and forces that do not exist. When they say they have found electromagnetic energy; that is meaningless because that means they have discovered light magnetism or electricity or some non- ionizing radiation which has nothing to do with weird goings on. There are only four forces in the universe, and we know them already. What are they?
The best answer is the NASA site:

"1. **Gravity** -—This force acts between all mass in the universe and it has infinite range.

2. **Electromagnetic** -—This acts between electrically charged particles. Electricity, magnetism, and light are all produced by this force and it also has infinite range.

3. **The Strong Force** -—This force binds neutrons and protons together in the cores of atoms and is a short range force. Strong Force holds the nucleus together which is made up of protons and neutrons which have quarks. Quarks are held together by the gluons.

4. **Weak Force** -— This causes Beta decay (the conversion of a neutron to a proton, an electron and an antineutrino) and various particles (the "strange" ones) are formed by strong interactions but decay via weak interactions (that's what's strange about "strangeness"). Like the strong force, the weak force is also short range. "

"The weak and electromagnetic interactions have been unified under electroweak theory (Sheldon Glashow, Steven Weinberg, and Mohammad Salaam was awarded the Nobel Prize for this in 1979). Grand unification theories attempt to treat both strong and electroweak interactions under the same mathematical structure; attempts to include gravitation in this picture have not yet been successful." [12]

http://imagine.gsfc.nasa.gov/docs/ask_astro/answers/980127c.html

---

[12] http://imagine.gsfc.nasa.gov/docs/ask_astro/answers/980127c.html

So when pseudoscience or religious individuals talk about forces or energies that cause miracles or supernatural behaviour, there are no other forces, just the ones discussed and that is a scientific fact.

32. What theory, in Quantum Physics, can give us teleportation, a perfect mail system and encryption for computers and quantum computers?

That is entanglement. Popular Mechanics on December 1, 2011 stated that the best way to entangle particles is to "entangle particles in the lab before, but only under special conditions, by isolating them and cooling them to ultra-low temperatures. The particles can then become entangled On a subatomic level; two particles are joined or correlated by a wave function (distance does not matter) that does not collapse until one of the particles are measured for spin (clockwise, anticlockwise) or polarization (vertical, horizontal).Neither particle has a definite state or form."
http://www.popularmechanics.com/technology/gadgets/news/spooky-quantum-entanglement-created-in-everyday-objects-6606439 . In fact, it has multiple states which we call quantum superposition. Once one of the particles is measured, the collapse of the wave function occurs, and the information of the spin or polarization is transferred as fast as the speed of the light to the other particle which has been classified as a Non-local. The many states of the particles stabilize to one each. E.g. If one particle is spinning upwards after being measured; the other suddenly starts spinning downwards. From this, we get the concept of non-locality, "spooky action at a distance". It described the ability of particles to become instantly aware of each other's state through an immediate transfer of information, even when separated by large distances, including if separated from different parts of the universe.

http://universe-review.ca/R13-11-QuantumComputing.htm Figure 03 Entan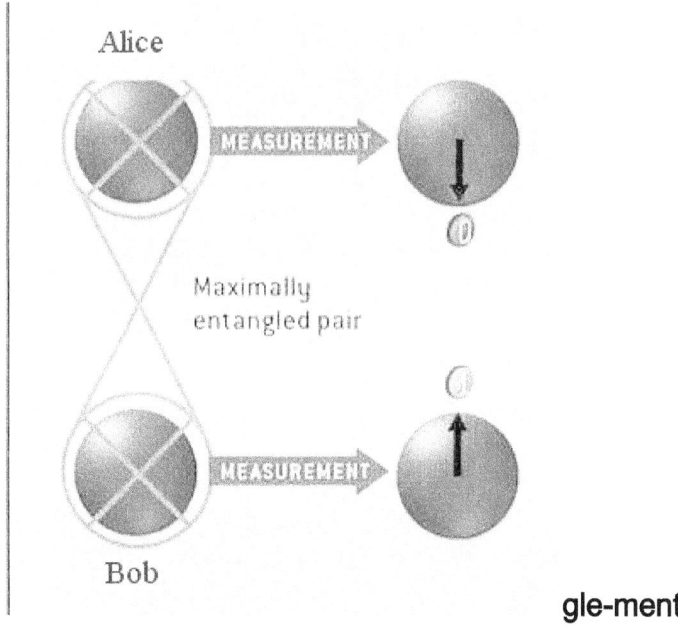gle-ment

33. How does entanglement work to help us?

Entanglement, in quantum computers, may create huge amounts of memory because qubits are in quantum superposition. They also produce low entropy, which is low disorder and results in less energy use by the computer. The non-locality (instantaneous transfer of the signal or state of information, to the other particle, when observed or measured that appears at almost the speed of light but is not) or the spookiness of entanglement will, in the future, contributes to theories of teleportation. Basically, non-locality is when each particle instantly becomes aware of each other's particle state, no matter how far away. I know many of you are hearing science definitions over and over again. Reinforcements will help you understand and comprehend concepts that are difficult. Scientists have created the first elementary quantum network based on interfaces between single atoms and photons using entanglement at The Max Plank Institute of Quantum Optics.

34. What is the Higgs Boson Particle and have we finally found it?

The Higgs has been known as the God Particle but not because it has religious significance. The name was formulated by Leon Lederman, a Nobel-Prize-winning physicist, for the title of a book called "The God Particle: If the Universe Is the Answer, What Is the Question?"

He was thought to have called it the "God damn particle"[13] but the publisher could not be used in that form.

Peter Higgs, an atheist, theorized the concept of the particle and was never happy about the name given by Lederman. It is the only Boson that we had not discovered, but we have some basic data now that it does exist. In the standard model, all particles are made up of Fermions, and Bosons and scientists believe subatomic particles gain their mass by interacting with the Higgs field, and the quantum energy field that fills all of space.

---

[13] http://www.forbes.com/sites/forbesleadershipforum/2013/10/09/the-higgs-boson-wins-the-nobel-why-we-call-it-the-god-particle/

The theory proposes that a so-called Higgs energy field exists everywhere in the universe. As particles move around in this field, some will interact with and attract Higgs bosons. Those particles will slow down and will attract greater amounts of Higgs bosons and the mass will be affected by the amount of Higgs bosons that are attracted to the particle going through the field. Others will go through the field unaffected like neutrinos and photons. There was the controversy that neutrinos could go faster than the speed of light, but that petered out when human error during the experiment occurred. Another simple analogy is that the Higgs energy field is filled with Higgs bosons and is like a sticky molasses, with Higgs bosons sticking to certain subatomic particles and thus allocating mass.

That is a little easier to explain at the mosh pit or the comic store. Higgs completed the Standard Model, and now we have much of what we need to go on to explain in terms of matter on earth. Discovery and knowledge, outside of our earthly boundaries, is another story. I think the best book on the Higgs is Sean Carroll`s book called Particle at the End of the Universe and the New York Times has a nice piece on the Higgs Boson.

http://topics.nytimes.com/top/reference/timestopics/subjects/h/higgs_boson/

At the Hadron Collider, in December, 2011, scientists studying data of millions of collisions thought they had spotted evidence that supported the existence of the particle. "We have narrowed down the region where the Higgs particle is most likely to be, and we see some interesting signals, but we need more data before we can reach any firm conclusions," 10 said Fabiola Gianotti, who headed the team that worked on the collider enormous Atlas detector. "We were expecting results in 2013 but something amazing was about to happen."[14] Evidence was also spotted at the Tevetron in 2011. However, on July 4, 2012, an announcement at CERN indicated that the high energy collisions had produced data of an energy mass of a particle resembling the Higgs, at 125Gev. The ramifications of this find will hopefully give us a connection to supersymettry and the verification of the standard model, an identity for dark matter, which may be WIMPS (**Weakly Interacting Massive Particles)** and a unification of the weak force and electromagnetic force. The discovery also communicates to the investors of the particle accelerators that it was worth all the money you gave to us. http://www.livescience.com/21381-higgs-boson-particle-implications.html . One last observation that should be noted; the matter that we are all familiar with is not really created by the Higgs Field. It is made from ordinary atoms. Brian Greene explains it best in his Facebook Page that was posted on July 16, 2012. "Ordinary matter is made from

---

[14] http://www.theguardian.com/science/2011/dec/13/higgs-boson-glimpsed-cern-scientists

atoms, whose mass mainly comes from protons and neutrons—which, in turn, are each made from three quarks. So while the Higgs gives mass to the quarks and other fundamental particles, it's the energy of the gluons that are responsible for most of the mass of the protons and neutrons, and hence the mass of familiar matter." The Higgs Boson type was verified at the beginning of September, 2012 and finally the actual Higgs on the week of March 18, 2013.

We are going to examine the terms Supersymettry and String Theory, so we can understand the ramifications of the Higgs discovery and look at the standard model of particles.

35. What is String Theory and why is it still a major contender for the Theory of Everything despite criticism?

The three major founders of String Theory were Yoichiro Nambu, Hoger Nielsen and Leonard Susskind and since then there have been many famous supporters. Each particle is composed of a vibrating string. A proton has three quark strings. When the quark strings come together, they form the proton. The strings are open, and some are closed and have been described looking like rubber bands and the vibration of the string determine the mass, type and charge. String Theory has evolved from Bosonic string theory to M-Theory. At this point, there are six different versions of String Theory. The main purpose of String Theory is to unify the four forces of nature with quantum mechanics. With the discovery of gravitational waves, some scientists think that we may be closer to supporting the theory of String Theory.

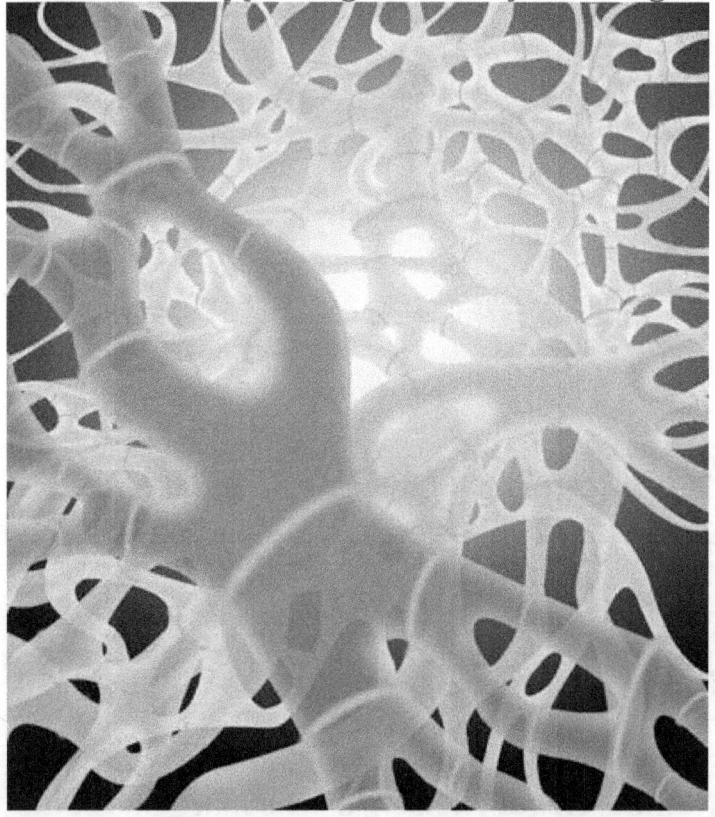

Artist`s impression of the fundamental entities of superstring theory by Flavio Robles Source: Berkeley Lab: http://www.lbl.gov/Publications/Currents/Archive/June-16-2000.html

)

   Those forces include weak nuclear force, strong nuclear force, relativity or gravity and electromagnetic force. I will show a table with all the particles that we know exists, but I will stress the major ones, so that you can remember them for posterity. Supersymettry is the reverse or superpartner particles of the Standard Model that only exist for String Theory. In the standard model, everything is made out Fermions and Bosons and then quarks and leptons under Fermions are quite popular.

Particles and Superpartners

Particle	Spin	Name	Feels These Forces[a]	Mediates These Forces[b]	Superpartner	Spin	SUGRA Mass Range [Atlas Lower Limit](Gev)
e, μ, τ	1/2	charged leptons (electron, muon, tau)	EM, W	---	sleptons $\tilde{e}, \tilde{\mu}, \tilde{\tau}$ (selectron, smuon, stau)	0	157 - 491, [$\tilde{\tau}$ > 136]
ν_e, ν_μ, ν_τ	1/2	neutrinos	W	---	sneutrinos $\tilde{\nu}_e, \tilde{\nu}_\mu, \tilde{\nu}_\tau$	0	230 - 486, [$\tilde{\nu}_\tau$ > 1320]
u, c, t	1/2	up, charm, top quarks	EM, W, S	---	squarks $\tilde{u}, \tilde{c}, \tilde{t}$	0	312 - 960, [$\tilde{t}$ > 310] [$\tilde{q}$ > 1400]
d, s, b	1/2	down, strange, bottom quarks	EM, W, S	---	squarks $\tilde{d}, \tilde{s}, \tilde{b}$	0	281 - 964, [$\tilde{b}$ > 390] [$\tilde{q}$ > 1400]
G	2	graviton	GR	GR	gravitino $\tilde{G}$	3/2	
$W^\pm$	1	weak boson	EM, W	W	Wino[d] $\tilde{W}^\pm$	1/2	97 - 741, [$\tilde{\chi}^\pm$ > 250]
Z	1	weak boson	W	W	Zino[d] $\tilde{Z}$	1/2	45 - 744, [$\tilde{\chi}^0_1$ > 300]
γ	1	photon	[c]	EM	photino[d] $\tilde{\gamma}$	1/2	45 - 744, [$\tilde{\chi}^0_1$ > 300]
g	1	gluon	S	S	gluino $\tilde{g}$	1/2	300 - 1009, [$\tilde{g}$ > 1770]
h	0	Higgs boson[e]	W	generates mass	higgsino[e] $\tilde{h}$	1/2	68 - 1024

[a] All particles feel the gravitational force.
[b] EM = electromagnetic force, W = weak force, S = strong force, GR = gravitational force.
[c] Photons feel only the gravitational force, but they interact with all electrically charged particles.
[d] Mixtures of these particles form charginos $\tilde{\chi}^\pm_1, \tilde{\chi}^\pm_2$, and neutralinos $\tilde{\chi}^0_1, \tilde{\chi}^0_2, \tilde{\chi}^0_3, \tilde{\chi}^0_4$.
[e] The additional Higgs bosons predicted by supersymmetry are not shown.

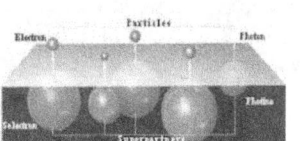

As Force Carriers –What they act on:

PROPERTIES OF THE INTERACTIONS

Property \ Interaction	Gravitational	Weak (Electroweak)	Electromagnetic	Strong fundamental	Strong Residual
Acts on:	Mass – Energy	Flavor	Electric Charge	Color Charge	See Residual Strong Interaction Note
Particles experiencing:	All	Quarks, Leptons	Electrically charged	Quarks, Gluons	Hadrons
Particles mediating:	Graviton (not yet observed)	W^+ W^- Z^0	γ	Gluons	Mesons
Strength relative to electromag for two u quarks at: 10^{-18} m	10^{-41}	0.8	1	25	Not applicable to quarks
3×10^{-17} m	10^{-41}	10^{-4}	1	60	
for two protons in nucleus	10^{-36}	10^{-7}	1	Not applicable to hadrons	20

©2004 -—2013 Astronomy Online.
The works within is licensed under a Creative

Commons Attribution-ShareAlike 3.0 Unported

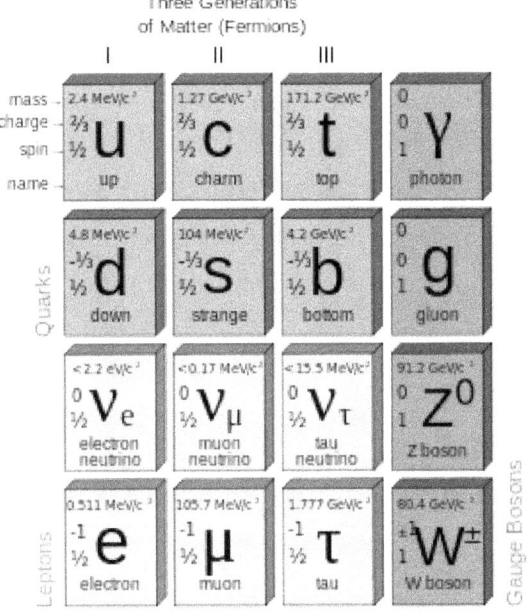

License

    Supersymettry is usually discussed in relation with String Theory which is sometimes called the Superstring Theory. Finding the super-partners would of the Standard Model go a long way to resolving many of the issues of String Theory. The superpartner of a fermion begins with an"s" and the superpartner of a boson ends in an "ino". I like the way Andrew Zimmer Jones explains it in his blog on August 12, 2012, in About Physics. "For each of the **fermions** (the leptons and quarks) there would be a counterpart that" a **boson**, and for each of the bosons (the force carriers) there would be a fermion counterpart. If the particle is a boson, then the superpartner is named by adding a ""ino""to the end. If supersymmetry is true, then the **photon** (a boson) should have a related photino which is a fermion.

The fermions, however, have superpartners which are named by putting an "s- prefix" onto the term. So the **electron** (a fermion) has a superpartner called a selectron, which is a boson." [15]

DO NOT GET THESE MIXED UP WITH THE ANTI MATTER PARTICLES FROM QUANTUM FIELD THEORY AND PAUL DIRAC AND VIRTUAL PARTICLES.

In String Theory, we are looking at many more dimensions; possibly 10 or 11, with the excess dimensions wrapped up in a beautiful object called a

Calabi-Yau Manifold

http://physics.about.com/od/physicsqtot/g/supersymmetry.htm

Source	Transferred from en.wikipedia; Transfer was stated to be made by User:Lunch.
Author	Original uploader was Lunch at en.wikipedia
Permission (Reusing this file)	CC-BY-SA-2.5.

---

[15] http://physics.about.com/od/physicsqtot/g/supersymmetry.htm

String Theory allows for Parallel Universes and the Multiverse . We will discuss this in another question. Joe Polchinski believed that Branes were part of the string theory. Branes are objects that resemble membranes. These membranes contain more dimensions that allow the attachment of strings and possible universes to be attached are speculated in this theory.

The holographic principal includes storing data on a 2d surface, which may reflect a 3d image, is fascinating because it can use the event horizon of a black hole or an area of the universe to do this. Leonard Susskind was a s supporter and a developer of the Holographic Principal. Edward Witten was responsible for the final drafts of M-theory as unification of superstring theories, beginning the second superstring revolution.

There are critics of String Theory, who accuse the theory of being difficult to test. These critics of String

Theory have alternatives. Some of those scientists include these opponents are also the same critics of the Multiverse. They include Lee Smolin and Roger Penrose. They are the Aeons Theory by Penrose describes a universe that reboots itself every aeon or cycle leaving behind CMB (Comic Microwave Background) and the Loop Gravity Theory by Smolin allows for the calculation of black hole entropy and an infinite singularity at the time of the big bang but no bang, a big bounce. Black holes produce universes within them. There is the Cyclic Universe by Paul Steinhardt of Princeton University and Neil Turok of Cambridge University. Neil is at the Perimeter Institute now in Ontario Canada. It is basically described as Brane (special membranes in space) collisions that produce universes. There is no Big Bang in the scenario. This is called an ekpyrotic universe.

You may want to research those theories but remember with the discovery of primordial gravitational waves; these theories are looking less probable. However, you never know until all the evidence is in.

Just so you as do not get mixed up with supersymettry, in Quantum Field Theory, the other world of antiparticles may be reinforced by this question.

36. What is the anti-world that resides next to us?

In 1928, Paul Dirac wrote down an equation, which combined quantum theory and special relativity, to explain the performance of the electron which won him a Nobel Prize in 1933. He discovered there is one for an electron with positive energy, and one electron with negative energy. Dirac interpreted this to mean that for every particle that exists, there is a corresponding antiparticle, exactly matching the particle but with opposite charge. For the electron, there should be an "anti-electron" identical in every way but with a positive electric charge. Dirac speculated on the existence of a completely new Universe made out of antimatter. Later scientists labeled the anti-electrons as a positron. Over the years, antiproton, anti-atoms and anti-neutrons were found. It was only in 2011 that CERN scientists were able to contain antimatter for sixteen minutes in the laboratory, in Switzerland.

We are watching the universe for planets, life and disasters with amazing technology.

37. Who kidnapped Pluto and will we pay the ransom?

http://stpaulcorrea.wikispaces.com/PLUTO

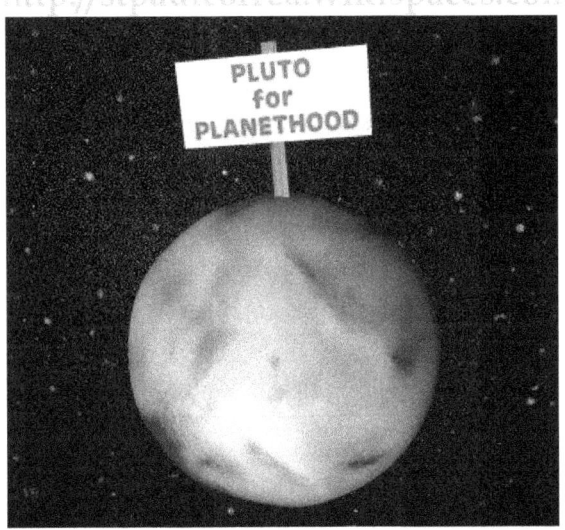

Clyde W. Tombaugh at the Lowell Observatory in Flagstaff Arizona discovered Pluto in 1930. Astronomers had always predicted there would be a ninth planet which they called Planet X. Tombaugh finally discovered an object, in a specific orbit, that he thought would be a match for Planet X and declared that he had discovered Planet X and the Lowell team was given permission to name it . They settled on Pluto, a name suggested by an 11-year old school girl in Oxford, England. For many years, the size, mass, and density of Pluto were thought to be similar to those of Mars. Then in 1978, James W. Christy discovered Pluto's satellite, Charon, using the 61inch astrometric reflector of the U.S. Naval Observatory in Flagstaff. Kepler's laws of motion were used to calculate the mass of Pluto and its mass was calculated as 1/400 of Earth's mass with a diameter of less than 2414km. The scientists realized that this data indicated that the planet Pluto was smaller than the moon. The International Astronomical Union (IAU) in Prague on August 24, 2006, decided to redefine the definition of a planet because of new sophisticated telescopes like the Hubble, were discovering new plantlike bodies, including Eris and Xena, and members of the Kuiper Belt were also revolving around the sun.

The new definition of a planet has three major properties.
They were as defined by the IAU
 "**1. It needs to be in orbit around the Sun-Correct**
2. It needs to have enough gravity to pull itself into a spherical shape – Correct

3. It needs to have cleared the neighbourhood" of its orbit"– Pluto is not a planet." [16]

What does "cleared its neighbourhood" mean? As planets form, they became the dominant gravitational body in their orbit in the Solar System. Pluto did not do this. This is defined by the IAU. Now Pluto is a dwarf planet which is assigned to planets failing the third criteria. Some scientists like Lawrence Krauss are not happy with the designation and others like Neil De Grasse Tyson encouraged the category change but still says that he was not responsible for it. This where we stand with Pluto. We now have eight planets. It looks like there is no ransom to pay, and Pluto has a new identity.

---

[16] http://www.iau.org/public_press/news/release/iau0603/questions_answers/

38. What telescopes, and technology, are searching the universe?

This is our solar system and most of these exoplanets (planets that orbit a star in the solar system other than that of Earth) that have been found by Kepler Project. What is it and are there other telescope projects?

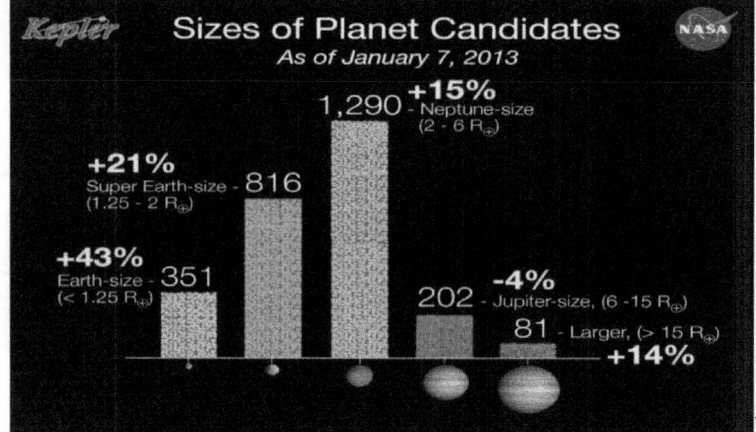
NASA PUBIC DOMAIN

Ball Aerospace designed and built Kepler's photometer and spacecraft and used its background knowledge of construction and expertise for the Hubble and Spitzer Space Telescopes. Kepler is the first NASA mission capable of finding Earth-size planets in the habitable zone. It was named after Johannes Kepler, who was a German mathematician and astronomer, who discovered that the Earth and planets travel about the sun in elliptical orbits. He denoted three fundamental laws of planetary motion. Kepler also did a study in optics and geometry.

Kepler Mission Object: (from the NASA site)

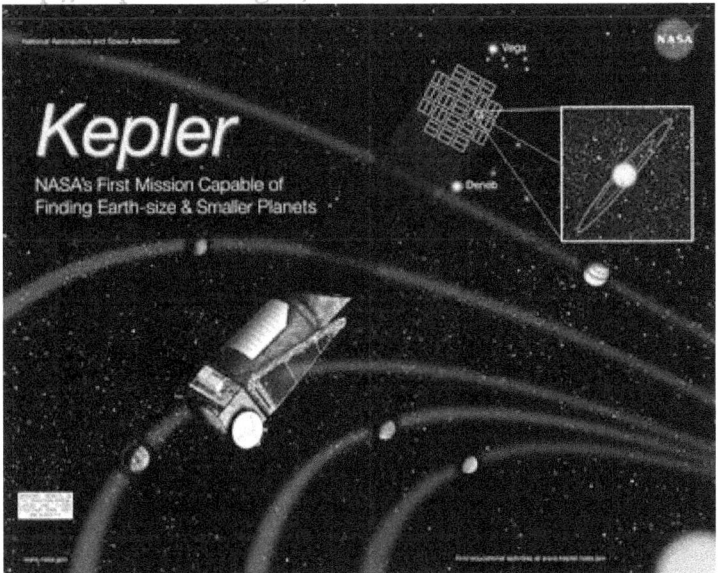

The scientific goal of the Kepler Mission is to explore the structure and diversity of planetary systems. This is achieved by surveying a large sample of stars to:

1. Determine the abundance of terrestrial and larger planets in or near the habitable zone of a wide variety of stars;
2. Determine the distribution of sizes and shapes of the orbits of these planets;
3. Estimate how many planets there are in multiple-star systems;
4. Determine the variety of orbit sizes and planet reflectivity's, sizes, masses and densities of short-period giant planets;

5. Identify additional members of each discovered planetary system using other techniques; and
6. Determine the properties of those stars that harbor planetary systems.[17]

It now has 1700 potential habitable planets. http://kepler.nasa.gov/news/nasakeplernews/index.cfm?FuseAction=ShowNews&NewsID=324

The Spitzer Space Telescope is a space-based infrared telescope studying objects ranging from our Solar System to the distant reaches of the Universe and still in operation and continues to look at Black Holes, Galaxies and exoplanets. Spitzer formerly called SIRTE (Space Infrared Telescope Facility) was the fourth and final element in NASA's Great Observatories Program. It was launched by a Delta rocket from Cape Canaveral, Florida on 25 August 2003.

---

[17] http://kepler.nasa.gov/

Source http://www.spitzer.caltech.edu/images/3072-SIRTF-Spitzer-Rendered-against-an-Infrared-100-Micron-Sky

Author NASA/JPL-Caltech

The NuStar Telescope was launched on June 13. 2012. NuStar may actually provide us with the first photos of a black hole rather than our depending on scrutinizing what surrounds the hole like gas and debris. From the site http://www.nustar.caltech.edu/about-nustar , The NuStar will be "probing cosmic ray origins to studying the extreme physics around collapsed stars to mapping micro flares on the surface of the Sun. NuSTAR will also respond to target of opportunity including supernovae and gamma-ray bursts."[18] NuSTAR (the Nuclear Spectroscopic Telescope Array) is a space-based X-ray telescope that will use unique lenses to focus high energy X-rays from astrophysical sources."

---

[18] http://www.nustar.caltech.edu/about-nustar

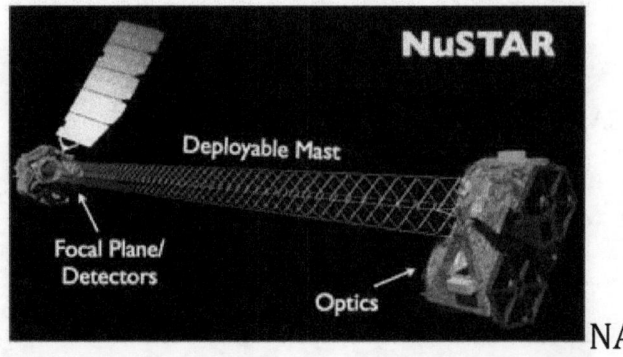
NASA

Chandra, the X-RAY-Ray Observatory, which was launched in July 23, 1999. It was a telescope that was formulated to discover X-ray emissions from very hot regions of the Universes such as, clusters of galaxies, and matter around black holes and exploding stars. It is still operating and functioning quite well. Photo: Chandra, NASA,

http://www.nasa.gov/vision/universe/starsgalaxies/chandra7years.html

The Hubble Telescope was the first major project. The Hubble telescope exploded to space aboard the space shuttle *Discovery* on April 24, 1990. It was named after Edwin Hubble, a brilliant American astronomer who expanded our knowledge of other galaxies and demonstrated that the universe is continually expanding through a cosmological constant. It has an eight foot mirror which had to be repaired after its initial first use in space. The Hubble is focused on examining dark energy, black holes, supernovas and each week the telescope sends back 120 gigabytes of data. It took a picture called "The Hubble Deep Field" which consisted of a detailed view of the universe. The image involved 342 different exposures taken over 10 days and illustrated some 1,500 different galaxies. It is the size of a school and orbits at the rate of five miles per second. It is still operating now and discovering new and challenging concepts within the cosmos. This will end in 2014. (Photo: NASA)

Finally on August 6, 2012, the Mars Rover called The Curiosity, landed in Gale Crater, to examine the planet for possibilities of microbes, water, methane, minerals, cosmic and surface radiation and find data useful for a manned flight to Mars. The landing was highly successful. You can go to http://www.nasa.gov/mission_pages/msl/index.html to see the amazing results of the discoveries and landing.

Source	http://photojournal.jpl.nasa.gov/catalog/PIA16239
Author	• NASA/JPL-Caltech/Malin Space Science Systems

NASA

39. How does a quantum computer differ from a normal computer and what can it be used for?

A conventional computer system operates on a digital operating system which involves a binary number system where there are only two numbers that being 0 and 1. Digital computers are fast and are useful for science, word processing, design and data processing. Quantum computers are not limited by 0 and 1 and incorporate their information on quantum bit or qubits. Qubits use the 0 and 1 simultaneously and are in a state of superposition, which we have discussed before. The quantum computers are probabilistic and random. The qubits have the ability to fluctuate between both states at once. Qubits act as computer memory and a processor through atoms, electrons and photons that are in a quantum event. The computer has then immense and untold power and memory that a traditional computer does not. This quantum event is entanglement which allows the qubits to fluctuate in superposition. The major problem is isolating the qubits, so they are not measured or observed to create a collapsing of the wave function. This action turns the computer into a normal digital one. They will be used to aid future technology in watching the universe with many SETI projects while offering simulations of solutions to many of earth's problems. This will also be our solution to Moore`s Law. Webopedia describes the Law as "The observation made in 1965 by Gordon Moore, co-founder of Intel, that the number of transistors, per square inch on integrated circuits, had doubled every year since the integrated circuit was invented. Moore predicted that this trend would continue for the foreseeable future. In subsequent years, the pace slowed down a bit, but data density has doubled approximately every 18 months, and this is the current definition of Moore's Law, which Moore himself has blessed. Most experts, including Moore himself, expect Moore'' Law to hold for at least another two decades."[19]

"Most experts, including Moore himself, expect Moore's Law to hold for at least another two decades."[20]

---

[19] http://www.webopedia.com/TERM/M/Moores_Law.html

[20] http://www.webopedia.com/TERM/M/Moores_Law.html

    I recall in my youth sitting in the planetarium listening to bands the thundering, metallic music of AC/DC and Pink Floyd as holographic, colorful images appeared everywhere through the simulated universe. Little did I know how close to reality this was in our universe, regarding the Holographic Principle?

40. Could our world or Universe be a hologram and are you really not kissing me?

The Holographic Principle evolved out of a conflict between Stephen Hawking and Leonard Susskind. Hawking hypothesized that anything or info that falls into a black hole is lost forever. Leonard took the holographic approach, in 1994, by Dutch Nobel laureate Gerard de Hooft. Hooft suggested "the entire universe can be seen as a two-dimensional information structure "painted "on the cosmological horizon." [21] In terms of a Black Hole, info would be in a 2D form and would be salvaged as a 3D Hologram in the flat surface of the Event Horizon. Hawking`s approach did not sit well with Susskind and he strenuously advocated that this would be a gross violation of the Law of Conservation of Information. Information cannot be lost. As I mentioned, on the surface of the black hole is a flat area called The Event Horizon. Hawking disagrees but we will leave his perspective because it is not tested or published. Susskind was aware that, if an object, let`s say satellite approaches a black hole or event horizon, he believed that the data or information about the object slowly spread over the entire surface area of the event horizon. The object appears stationary or intact just above the surface, and the information spreads all over the 2d surface. However, the satellite simply falls through the horizon into the inescapable pull of the black hole but as it does a 3d image is emitted with the radiation, which is called Hawking`s Radiation. Hawking eventually gave way to Susskind`s Theory which is outlined in a great book called, "A The Black Hole War by Leonard Susskind".

---

[21] http://www.sciencedaily.com/articles/h/holographic_principle.htm

The strange theory that we see is nothing more than the 3-d representation of information encoded on the 2-d surface of the event horizon has evolved from the holographic principle. It had been speculated that this theory can also be applied to a part of our universe which holds all our information on a flat 2d area emitting a 3d version of this universe. We have an instrument called the GEO600 which is looking for holographic noise to support this theory. Now there is controversy about what happens to this information again. Some scientists have put forth a theory about "firewalls" which destroy everything entering the black hole. Hawking put forth an explanation to counteract this theory by eliminating event horizons.
http://www.universetoday.com/108870/why-hawking-is-wrong-about-black-holes/

I am again going to answer this as a possible yes and repeat the two major links that have occurred, early 2013 in December, in Nature Magazine and a rebuttal. They are
http://www.nature.com/news/simulations-back-up-theory-that-universe-is-a-hologram-1.14328
and http://www.universetoday.com/107172/why-our-universe-is-not-a-hologram/   If you get a chance, check out Ethan Siegel`s Twitter site, for more information, on this issue, or you can engage in enjoying his blog at
http://scienceblogs.com/startswithabang/ .

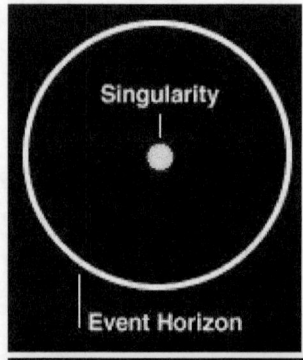

http://phantasticphysics.wikispaces.com/+Event+Horizon+of+A+Black+Hole

41. What are holograms used for now? (See Picture display)

Holograms are being used for music stars like Tupac and Caochella and possibly Elvis in the future. They are used for advertising and simulation experiences in areas of medicine, military, credit cards and the ecological displays

Author

Jase The Bass Band: X Japan.

Source http://www.flickr.com/photos/jasethebass/3209854164/in/set-72157594266410289/

Author Jase The Bass

Source http://www.flickr.com/photos/jasethebass/3209854164/in/set-72157594266410289/

Author Jase The Bass Band: X Japan.

42. Is there life in the Universe?

I will never forget watching Forbidden Planet (1956, director. Fred M. Wilcox). It is a science fiction classic that is said to be the first movie to take place on another planet, in our universe. I was in my early teens .It was supposed to have been modeled as a retelling of Shakespeare`s The Tempest, but Forbidden Planet had the most amazing, special effects with a monster that comes from underground to wipe out individuals on a lonely

outpost.

Origem http://wrongsideoftheart.com/wp-content/gallery/posters-f/forbidden_planet_poster_01.jpg

Author   Copyrighted by Loew International.
         Artists(s) not known.

Public Domain.
Special effects here are exceptional.

This film examines our own experience and the human race self-destructiveness. As I sat there watching this incredible movie, I wondered if there were other planets and solar systems that had life and would there ever be a way of discovering how many did without visiting them. I had no idea that a brilliant scientist named Frank Drake, who today is employed by the, Search for Extraterrestrial Intelligence Institute, had created a complex formula that would give us the potential number of extraterrestrial civilizations in our galaxy and it is used by institutions and scientific organizations all over the world today. SETI actually states on its website (http://www.seti.org/about-us) that "The mission of the SETI Institute is to explore, understand and explain the origin, nature and prevalence of life in the universe." The SETI Institute was organized on February 1, 1985. It employs over 150 staff members including scientists, media, and research and support staff. They are sponsored by NASA and such huge organization such as Hewlett Packard and Sun Microsystems while at the same time enjoying a nonprofit status with the U.S government. They use everything from Astrobiology Telescopes and Satellites to try to find evidence of life. Frank Drake works as a radio astronomer at the National Radio Astronomy Observatory in Green Bank, West Virginia, and currently sits on the Board of the SETI Institute. Drake initiated the first radio search for civilizations outside the limits of Earth, and he assisted Carl Sagan design a plaque that was connected to the Pioneer 10 and 11 spacecraft which the beginning of humanity to attempt to communicate outside the Solar System. Even though nothing has been found, in terms of alien life, Drake believes we are close as technology expands. What motivated him and the SETI movement

was his equation called the DRAKE Equation. Scientists now input the latest data and factors and come up with the possible alien like civilizations. Here is how it is written. The equation is usually written:
$N = R_* \cdot f_p \cdot n_e \cdot f_\ell \cdot f_i \cdot f_c \cdot L$ Where,
N = the number of civilizations in The Milky Way Galaxy whose electromagnetic emissions is detectable.
R* =The rate of formation of stars suitable for the development of intelligent life.
fp = The fraction of those stars with planetary systems.
ne = The number of planets, per solar system, with an environment suitable for life.
fl = The fraction of suitable planets on which life actually appears.
fi = The fraction of life bearing planets on which intelligent life emerges.
fc = The fraction of civilizations that develop a technology that releases detectable signs of their existence into space.
L = The length of time such civilization
http://www.pbs.org/lifebeyondearth/listening/drake.html

In an interview in Cosmos, February 2010, Frank Drank stated that he believed based on his formula, that the number of planets with civilizations just in our galaxy is probably around `10,000. .[22] SETI has software that allows you to participate in discovering extraterrestrial signals right in your own home. You could be the one in the history of our human existence

---

[22]   http://www.cosmosmagazine.com/planets-galaxies/qa-with-frank-drake/

to discover ET. Go to this site.
http://setiathome.berkeley.edu/

Sara Seager, professor of planetary science and physics at the Massachusetts Institute of Technology has updated the formula this in 2013. Where:

- N is the number of planets with detectable biosignature gases
- N* is the number of stars within the sample
- F_Q is the fraction of quiet stars
- F_{HZ} is the fraction with rocky planets in the habitable zone
- F_O is the fraction of observable systems
- F_L is the fraction with life
- F_S is the fraction with detectable spectroscopic signatures[23]

---

[23] http://io9.com/what-a-brand-new-equation-reveals-about-our-odds-of-fin-531575395

Check out this article:

Photo Seti

http://www.gravita-zero.org/2009/10/non-probabilita-di-vita-extraterrestre.html

As a teenager, there was nothing I wanted more to do than to travel back and forth in time. I still live vicariously through Dr. Who and his adventures.

43. Is time travel possible?

Attribution-**NonCommercial-ShareAlike 2.0 Generic** (CC BY-NC-SA 2.0)

Image: **Corvin Zahn**, Institute of Physics, Universität Hildesheim,**Space Time Travel** (http://www.spacetimetravel.org/)

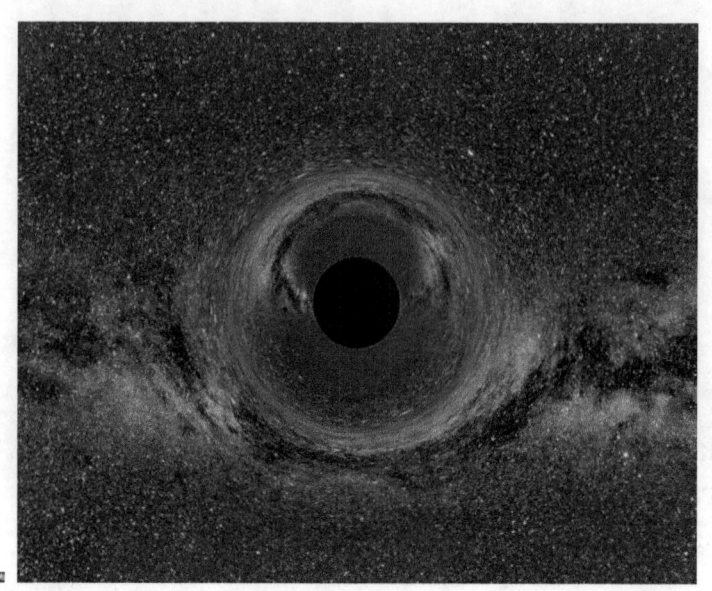

Credit: NASA CD-98-76634 by Les Bossinas.

Einstein`s theory of Relativity never had any problem with Time Travel to the future. Where difficulty arose was whether one could go backwards in time and whether relativity itself did allow for that process. Let`s examine how time travel to the future could occur under present laws of physics and how it was presented in Comics and Science Fiction. Superman has had the ability to fly break the Law of Relativity by flying faster than the speed of light, thus allowing him to go in reverse backwards in time. However, at this point if Superman were real, he could not really go backwards in time because the laws of physics would not allow him to go faster than the speed of light.

dragonofbrainstorms

Creative Commons Attribution-Share Alike 3.0 License

Superboy by man0ftomorrow
http://man0ftomorrow.deviantart.com/art/Superboy-373283720

Creative Commons Attribution-No Derivative Works Licenses.

Let us look at two time travel stories The Impossible Mission" from Superboy *#85* (1960) and "Rip Van Superman" from Superman #107 (1956). In the Impossible Mission, Superboy travels back in time to stop the assassination of Abraham Lincoln and when he finds out Lincoln is staying at a hotel before going to the Ford Theater; he plans to warn him. When he gets there, he finds Lex Luthor has signed in under the same name as Lincoln. Mr. L. Luthor thinks the adult Superman has sent Superboy to defeat him and exposes Superboy to Red Kryptonite, which almost kills Superboy. Suddenly Luthor hears gunfire and realizes that Superboy meant to stop John Wilkes Booth and had no idea that Luthor was in this time zone. You cannot change time. In Rip Van Superman, A Kryptonite cyclotron accident puts Superman in a coma. He is placed in a tomb, and sleeps until the $30t^h$ Century. When Superman arises from a deep sleep, he learns that drugs have given all Earth citizens powers comparable to his own. The Earth population is using cyber machines to complete their work, and Superboy gets himself into a confrontation with the villain named Drago. Superman organizes the people of the $30t^h$ Century to fight Drago, and they are victors. Superman, as a result of victory, is sent back to his present time by the keepers of time with an experimental time travel machine. He awakens in his glass tomb only minutes after the initial placement in the tomb. In this scenario, time travel has a profound effect on the future.

In Science Fiction, one of the most successful of the time travel characters is Dr. Who and he has been around on TV and in books for almost fifty years. This is the fifty year anniversary of Dr. Who with the humungous release of the Dr. Who movie called Day of the Doctor. http://www.bbcamerica.com/doctor-who/extras/takeover-schedule/.

This was broadcasted all over the world in most countries from Canada, U.S, France and 84 countries. The celebration even included a Google Logo and doodle with an eight bit game. http://mashable.com/2013/11/21/doctor-who-google-doodle/

Dr. Who travels through the past, present and future of the universe with his cool companions fighting murderous entities, robots, aliens while also saving alien civilizations and negotiating information about his own secretive history. I have lived vicariously through the imaginative and creative episodes that started with primitive special effects in the early series to the digital computerized technology used now. The time travel all takes place on the Tardis, which is an old fashioned English Telephone Booth of the Fifties. The Tardis stands for Time and Relative Dimension in Space, and the Doctor indicates that the Time Machine is indestructible because of the force field surrounding it. Dr. Who accepts the classical view of the universe with only three dimensions with the fourth being time. It integrates the three primary dimensions of the universe while only the one part, the engineering and operating room, materialize when it lands. It was, therefore, bigger on the inside than on the outside. The

Time Machine runs on massive energy from a captive black hole (which of course violates the laws of physics) called The Eye of Harmony. The solution to paradoxes and going backwards in time is the creation of alternative timelines for the Tardis .Thusly eliminating the paradox of actually meeting yourself.

In Borrowed Time, an Interview with Michio Kaku, a theoretical physicist, who is Dr. Kaku, contemplates the plausibility of time-travel in the article by J.R. Minkel in Scientific American November 23, 2003. Dr. Kaku states that if wormholes were found to exist, we could utilize one to travel forward or backwards. He says, "Wormholes, from the vacuum, would connect randomly with any point in space and time, so the other end would connect God-knows where. Probability-wise, the wormhole would be more likely to connect with the Universe back in time rather than the Present." [24]

Many scientists have changed their mind about time travel with wormholes because of stability issues in the wormhole. Any changes that were to occur because of a quantum paradox situation would result in a parallel world with a parallel time line (where the paradox did not exist).This being created by means of the Many Worlds' Theory by Hugh Everett. This is similar to what the Tardis writers have used to avoid paradoxes in their books and series. Let's look at time travel in terms of the evidence and information we have now and the theories that seem to have a lot support at this point.

---

[24] http://www.scientificamerican.com/article.cfm?id=borrowed-time-interview-w

Magnus D.
Author ⓘ

Relativity, by Einstein states that the closer you go to the speed of light than, the slower time gets, and this is called time dilation. If I am in a rocket ship from Earth to Planet A and I am going very close to the speed of light, relativity indicates that time dilation will occur. I may age two years, one way to Planet A and then another two years back to Earth from the planet A, so that I personally have aged and gone four years into my future. However, when I get back to Earth, I find that the time on Earth has progressed fifty years, and I have visited the Earth 46 years into the future. Once scientists create the technology for travelling vast distances where there are black holes, astronauts will be able to manipulate those black holes also for time dilation, as well. A black hole bends time and space creating a gravitational attraction where nothing can escape not even light and as you get closer the time dilation get stronger. You would have to stop at a point before the event horizon so you would not get sucked in but when you did stop and hover, you would get two to one time ratio as in five years becomes ten. The black hole may provide another useful approach to time travel. We can use this for travelling to the future, and we can also use black holes and wormholes (when and if they are proven to exist and they are stable) to further our ability to time travel in the future. Wormholes surround one end of the black hole.

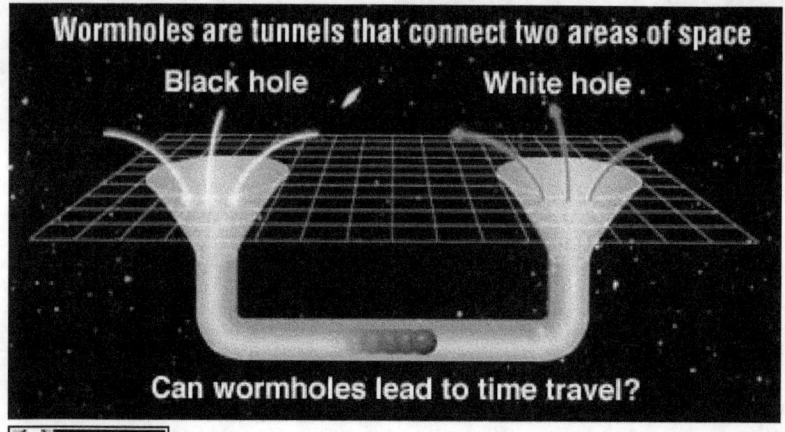

http://306physics.wikispaces.com/Wormholes

There is no real hard evidence for wormholes or what is called an Einstein-Rosen Bridge, which encompasses two points in space-time when connected. We would want to find a way of connecting, using this bridge, two regions of different times. A man named Ludwig Flamm had theorized this idea in 1916 and Einstein and Rosen studied this concept in the 1930's. The name "wormhole" was coined by American theoretical physicist John A. Wheeler in 1957, based on how a worm could eat a hole from one end of an apple through the center to the other end as a result producing a "shortcut" through the universe. Wheeler also realized that the wormhole would be totally unstable and immediately collapse but physicists Kip

Thorne and Mike Morris suggested since that such a wormhole could be made stable by introducing some form of negative energy. Robert Hjellming, in 1971, described a wormhole model in which a black hole would extract matter, while joined to a white hole at a distant location, which throws out this same matter. Inside this wormhole, you could take a rocket ship from point A to B, starting in 2012, at one end of the bridge and ending through time dilation at 4089 at the other end.

   In String Theory for Dummies, Andrew Zimmerman Jones talks about J. Richard Gott and William Hiscott who, in 1991, thought that the mathematical theory of cosmic strings (which predated String Theory) might be a great way to time travel. Now scientists believe Gott`s versions of cosmic strings are close in definition to their version of String Theory, and if we finally prove String Theory to be a Theory of Everything, we may have a new form of Time Travel. A lot of ifs but remember, these theories were never pulled out of hats but were proposed with mathematical and suggestive evidence that solved problems on a theoretical basis. Here is how it works. Find a black hole with its singularity (a tiny spec in the space-time continuum where the black hole began). Fire two or more cosmic strings, close to the speed of light, around the black hole and create a whole array of closed time like curves around the black hole. Then move your ship in a figure eight positions around the strings transcending the future and present. This is not for our present technology. We talk about future, past and present, but really physics says the direction of time is just a perception and not real.

http://www.epubbud.com/read.php?g=4T N6PFQ3&107 is from Andrew's book where these concepts are discussed.

We must also remember when we build a time machine for the first time; this is as far as we can go back in time. So do not expect to see dinosaurs or caveman. Here are the various time travel rules by Sean Carroll. http://discovermagazine.com/2010/mar/02-the-real-rules-for-time-travelers#.Up4eKPtFPRg I also recommend "Time Travel for Beginners", by John Gribbin.

44. What are your favorite movies or TV episodes using antimatter?

We shall remember that scientists have only been able to hold the antimatter for about 16 minutes before it pops out of existence. It was produced at CERN's antimatter factory, more formally known as the Antiproton Decelerator. In ordinary atoms, negatively charged electrons, move around positively charged protons. When atoms of matter and antimatter come into contact, they annihilate each other and everything in the vicinity. One of my favorite movies for antimatter is Angels and Demons, by Dan Brown which is a follow up to the novel The Da Vinci Code. This is where a vial of antimatter is stolen to create a massive bomb to destroy the Vatican. Of course in real life antimatter would disappear immediately from the vial. It is a very rare commodity, so there would never be enough located at the CERN to create anything resembling a bomb. CC
http://wiki.creativecommons.org/Case_Studies/CERN
Star Trek was powered by antimatter propulsion systems.

A great Warp Core Breach is from Star Trek Voyager Waking Moments
http://www.youtube.com/watch?v=joiXzDoBjGk.

   Before all these amazing particle discoveries can occur, we must have particle accelerators that provide high energy collisions.

45. What are the most famous Particle Accelerators?

The Large Hadron Collider (LHC) is a particle accelerator created by CERN (European Organization for Nuclear Research).   It is a particle physics laboratory based in Geneva, Switzerland. It is the world`s largest organization devoted to particle physics. A particle accelerator sometimes called an "atom smasher" (http://www.youtube.com/watch?v=1sidBwpv0Zg) by lay people, and it is a device that propels subatomic particles called hadrons at high speeds. The Collider has examined such concepts as the Higgs Particle, the speed of Neutrinos, dark matter and energy, and antimatter.   Fermilab `s famous accelerator is called the Tevatron, but it was closed down in 2011. However, Fermilab is now planning a Linear Accelerator and the rest of Fermilab's chain of accelerators will be replaced with a more efficient particle source in 2014. They also had been looking for the Higgs but even though the Tevatron is nonoperational, they still had analyzed a huge amount of data which contributed to the discovery of the Higgs.

Here is a list of all the Particle Accelerators in the World, look them up if you really cannot find anything to watch during summer reruns.
http://www-elsa.physik.uni-bonn.de/accelerator_list.html

46. What is the scientific method and why is it important?

The scientific method is a way of discovering the truth in a natural and material world using empirical evidence. There is no evidence that there is anything other than the material and natural world, no matter what anyone tells you; therefore, the only way to test anything is through hypothetical-deduction-ism or the scientific method. When we reason out problems, most of us do it intuitively but that is the lowest form of analysis. The highest for is through analytical and critical learning skills which go hand in hand with the scientific method. Nerds follow science and skepticism. Here is how the scientific method works:

Simply put collect and observe data, hypothesis, experiment, results, verify, reproduce. Peer review,

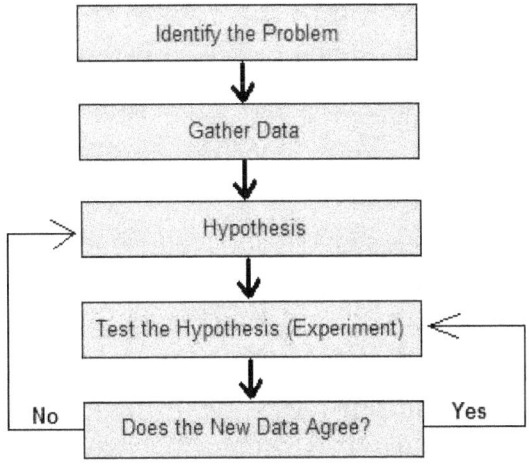

Publish.

Source File:High_School_Chemistry.pdf, page 23

Author CK-12 Foundation

We can modify this to apply it to your life, as a worldview, so that you use it as a way to examine everything, from religion to the cost of the battery, at the car dealership. Scientists and people who use this method as part of a world view tend to lose interest in religion and God.
http://news.discovery.com/human/religious-belief-critical-thinking-120426.html and http://www.scientificamerican.com/article.cfm?id=how-critical-thinkers-lose-their-fa . Also, declarations by Hawking and Krauss have led both of them also to suggest a universe, self-sustaining, without the supernatural and God, based on the scientific method being used to support theories (which have been over sixty years in the making and supported and contributed to by other scientists). . In March 2012, New Scientist indicated that 93 percent of the members had no personal God. "93 per cent of the members of the academy do not believe in a personal god." [25] This is a 1998 study before the actual resurgence of Atheism and the huge dropping off conventional religious membership in the between 2010 and 2013. This shocked America this summer from a CBS poll.
http://washington.cbslocal.com/2013/07/02/study-nearly-half-of-americans-say-increasing-non-religious-sect-troubling-for-society/

25

http://www.newscientist.com/article/mg21328562.300-the-god-issue-god-is-a-testable-hypothesis.html Victor J. Stenger, The God Issue

47. Nerd Guide to Famous Directors and their best movies and television programs. My top choices have photos.

A.J.J Abrams

Person of Interest
Star Trek
Mission Impossible

Alias
http://series.tvfilia.com/2008/11/alias-online-capitulos-completos.html
Fringe
Lost

If you think Lost was good after that ending, read the section on critical thinking.

B. Ridley Scott
Alien

Source   Alien from AVP

- Uploaded by MaybeMaybeMaybe

Author   Cristiano Betta from London, UK

Blade Runner
Good Wife
Thelma and Louise
Prometheus –Actually screwed up by the same writer of Lost. I will not mention his name.

C. CHRISTOPHER NOLAN

Memento

James **Digital_Third_Eye   Brooding Batman
Attribution-NoDerivs 2.0 Generic** (CC BY-ND 2.0)

Inception

D. Guillermo Del Toro

Pans Labyrinth

Brett Jordan   Attribution 2.0 Generic (CC BY 2.0)

E. Josh Whedon

http://teachingfilmandmedia.wikispaces.com/Gender+Portrayals+in+Pop+Culture

Buffy the Vampire Slayer

Angel

The Avengers
Cabin in the Woods
Agents of Shield

F. George Lucas
Star Wars Series I-III

George Lucas CC BY 2.0 view terms
http://creativecommons.org/licenses/by/2.0/
Joi Ito from Inbamura, Japan - Taken from **Flickr**: **link to original description page**

G. Steven Spielberg

E.T   CC
Public Domain
Schindler`s List

Pinky`s Brain
Who Framed Roger Rabbit
Jurassic Park

Who Framed Roger Rabbit logo Public Domain

Walt Disney Pictures, Amblin Entertainment - movieposterdb.com; rotated by 270

H. Jon Favreau   Iron Man

http://www.coestrenos.com/rel/ironman

I. James Cameron

The Terminator

Some rights reserved by Yortw Avatar License

J. John Carpenter

Halloween Some rights reserved by TRF_Mr_Hyde

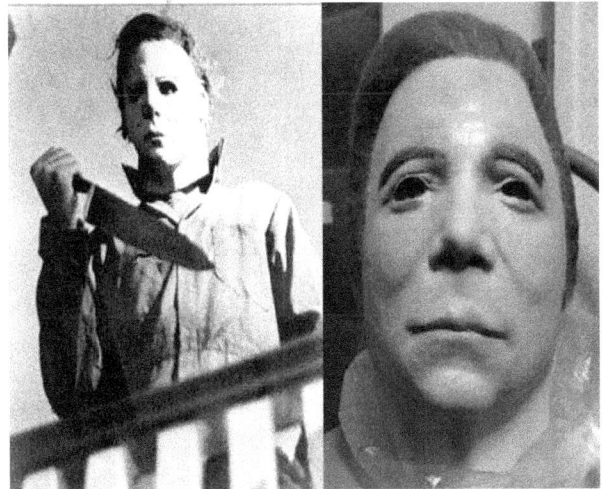

K. Stanley Kubrick

by kainita on Flickr
Some rights reserved

2001 A Space Odyssey

Some rights reserved by x-ray delta one

L. George A. Romero

Night of the Living Dead Orig.

License
ⓘ ⊜ **Some rights reserved** by Jesper Wiking

M. Andy Warchowski

Attribution 2.0 Generic (CC BY 2.0)

The Matrix Teaser Poster by TinyButDeadly
The SOME RIGHTS RESERVED. This work is licensed under a
Creative Commons Attribution-No Derivative Works 3.0 License.

N. William Cameron Menzies
Invaders from Mars

Some rights reserved CC Author: Adolfo Suarez

D. Joe Dante

Gremlins Inti
Some rights reserved **Attribution 2.0 Generic**
(CC BY 2.0)

P. Steven Moffat

Sherlock

http://samuraifrog.blogspot.ca/2010_12_26_archive.html

Best Nerd Sci-Fi TV Shows
Dr. Who

Creative Commons Attribution-Share Alike 3.0
©(attribution & link required)
See original image page
By: alun.vega
Source: Flickr

Fringe

Public Domain
http://commons.wikimedia.org/wiki/File:Logo_Fringe.jpg

True Blood

TRUEBLOOD

http://commons.wikimedia.org/wiki/File:True_Blood_2008_logo.svg Public Domain

Torchwood No. 2 ⓘ ⊜Some rights reserved by alacoolb

Best Classic Sci-Fi

The Wild Wild West; I give it to this one.

http://thetvdb.com/?tab=series&id=71637
Superman
Smallville
Alf
Twilight Zone
Batman

4B. Scientist and Individuals that you must look up and discover in the Nerd World.

These are the scientists and amazing writers that you will understand and learn a great deal from after reading my book and if you want a peak at the Nerd Universe of Knowledge. These scientists and writers are amazing communicators willing to share a wealth of information, in all areas of scientific disciplines. You will enjoy their books immensely and posts. Your life will be changed by them.

Stephen Hawking –

Thinks the M-Theory is the unified theory that Einstein was looking for. Hawking was diagnosed with Amyotrophic Lateral Sclerosis (ALS). Despite his devastating illness, he has done incredible work in physics and cosmology that has lead to discoveries that include Hawking`s radiation and new insights into relativity and his several books try to make science accessible to everyone. He is one of the world's most brilliant minds. The book was written with

Leonard Mlodinow

; I recommend reading "The Grand Design".

Mr. Mlodinow is another brilliant physicist who has written a brilliant book about the randomness about the universe called "The Drunkards Walk" and the "War of Worldviews".

John Gribbin.

He is an incredible astrophysicist who has written books on music, climate change, quantum physics, evolution and other famous scientists. My favourite books are "A Computing with Quantum Cats: From Colossus to Qubits" to "Alone in the Universe" and "In Search of Schrödinger Cat: Quantum Physics and Reality". This man is one of the great science communicators of all time. He was one of my inspirations for writing this book. Thank you, John.

Sean Carroll

He is an incredible cosmologist especially in quantum physics, dark energy and general relativity. His Cosmic Variance blog is one of the best on the internet and his book discussing Entropy and the direction of time, called "From Eternity to Here", and should not be missed. What an Amazing book. His new book on the Higgs is worth picking up, as well. It is called "The Particle at the End of the Universe". He and his wife were inspirations to my science hobbies and writing the book.

Tommy Rodriguez,

is a Wildlife enthusiast, biologist and amateur photographer, who has one of the best books on the wildlife of the Everglades. It is called Visions of the Everglades: History Ecology Preservation. His studies on genetic biology are incredible. Learn about evolution and cosmology at http://www.tommyrodriguez.com/ **and** http://pangaeabio.com/?page_id=2234

Carl Zimmer

Carl Zimmer is one of the best science writers in the world today. I recommend His books "Planet of the Viruses" and "Science Ink: Tattoos of the Science Obsessed". www.carlzimmer.com

Peter Boghossian

Peter is a one of the top secular philosophers today discussing religion, faith and science. This is from his biography in Wikopedia. "Dr. Peter Boghossian's focus is bringing the tools of professional philosophers to people in a wide variety of contexts. Peter has a teaching pedigree spanning more than 20 years and 30 thousand students in prisons, hospitals, public and private schools, seminaries, colleges and universities, Fortune 100 companies and small businesses. His fundamental objective is to teach people how to think through what often seems to be intractable problems. " He teaches at Portland State University. His latest and greatest book is "A Manual for Creating Atheists".

Guy P. Harrison

is one of the top skeptics and science authors in the North America and he has written one of the most entertaining and informative books on the market called "50 Popular Beliefs that People Think are true". Other excellent books include "Race and Reality" and "50 Reasons Why People believe in a God". Guy has held numerous positions in the news industry, including editorial writer, world news editor, sports editor, photographer, page designer, and columnist. His new book "Think" is his best book that examines pseudoscience and skepticism.

Brian Greene is probably the first physicist that exposed me to the concepts of the mulitverse. His main theory of expertise is String Theory and he hopes to demonstrate that it will unify the four forces of the universe. It will be the Theory of Everything. I recommend "The Hidden Reality", "The Elegant Universe", "The Fabric of the Cosmos" and an essay called Welcome to the Multiverse. Also try to watch the DVD versions of his books. You can contact him **at physics@briangreene.org**

Lawrence Krauss

When discussing Dr. Krauss, it is best to outline from his own biography on Wikipedia. "Is an American theoretical physicist and cosmologist who is Foundation Professor of the School of Earth and Space Exploration at Arizona State University and director of its Origins Project. He is known as an advocate of the public understanding of science, of public policy based on sound empirical data, of skepticism and of science education. He is also the author of several bestselling books, including "The Physics of Star Trek" and "A Universe from Nothing". "[26] Dr. Krauss has been an inspiration for me in the area of quantum physics and atheism. He has taken the works of Alan Guth to Stephen Hawking and others and has taken them in a new direction and clarified their theories so that non-scientists can understand what the major theories are in terms of how the universe can be created spontaneously without the "shenanigans"[27] of the supernatural or God. I recommend A Universe from Nothing and Quantum Man.

---

[26] http://en.wikipedia.org/wiki/Lawrence_M._Krauss

[27] http://www.ttbook.org/book/transcript/transcript-lawrence-krauss-marcelo-gleiser-something-nothing

Nicola Tesla

Nicola Tesla was born in the Hungarian **Austrian Empire, in** 1856, in a place called Lika. Many think his scientific achievements have been underestimated. He discovered the induction motor and developed a plan for the AC Motor when he was in his twenties, in New York City. Thomas Edison hired him to help fix his DC Motor but refused to pay him when it was completed. Tesla`s AC Motor eventually became popular but was always competing against Edison. Then he developed patents for coils that would transmit radio signals, but these ended competing against Guglielmo Marconi`s later patents. Marconi won even though Tesla was first. Tesla really had the upper hand in many areas from lighting to remote controls. Finally Niagara Falls was to incorporate his AC Motor Design with 5000 horsepower design by Westinghouse. He was the ultimate Nerd who was never appreciated for all he accomplished.

http://www.pbs.org/tesla/ins/index.html

Seth Lloyd

A great book on how the Universe may be a quantum computer and how quantum computers operate and function. The book is called Programming the Universe.

Henry Gee

He is the senior editor of Nature Magazine and has provided me with much of my knowledge of Evolution with one of the best books written on Evolution called "The Accidental Species". He is a brilliant and nerdy scientist who is an expert on Tolkien and old school rock and metal. I totally absorb everything that is written by this individual.

If you want to know about the scientific fallacies and old wives tales of life, brilliant science writer **Dave Bradley**'s new book "Deceived Wisdom", is for you. He is a well renowned and respected science writer, who will provide you with constructive educational knowledge in many disciplines, on his website Sciencebase. http://www.sciencebase.com/science-blog/.

If you are a real nerd, you must visit the Perimeter Institute. To quote Wikipedia," Perimeter Institute for Theoretical Physics is an independent research centre in foundational, theoretical physics located in Waterloo, Ontario, Canada. Founded in 1999, Perimeter Institute works to advance understanding of physical laws and develop new ideas about space, time, matter, and information."

Thank you to Neil De Grasse Tyson, who is an astrophysicist and science writer and the director of the Hayden Planetarium. He just finished hosting the new Cosmos Series, which is a reboot of the old one by Carl Sagan. He has appeared on shows from NOVA, Bill Maher to The Daily Show. His amazing and passionate speeches for space travel and discovering our universe is inspiring. I recommend his book Space Chronicles: Facing the Ultimate Frontier. Cosmos began on Fox and the National Geographic channel with rave reviews and homage to Carl Sagan.

(cc) BY-SA
http://www.cosmonoticias.org/category/otros/page/

Carl Sagan:

Carl was probably science's best communicator when it came to transmitting cosmological concepts and outlining the process of critical thinking, and its application in discerning the non validity of the supernatural and religion. He wrote over a dozen books, worked for NASA and was a regular on The Tonight Show with Johnny Carson. He had the original series called Cosmos, on PBS, in thirteen parts. He admitted his agnostic feelings to skeptic Martin Gardener, criticizing Martin's belief in God, Bible and the afterlife. From the Smithsonian article; "In November 1996, Sagan wrote to Gardner: "The only reason for this position that I can find is that it feels good….How could you of all people advocate a position because it's emotionally satisfying, rather than demand rigorous standards of evidence even if they lead to a position that is emotionally distasteful?" [28] He died in Dec. 20th, 1996 and left his brilliant wife Ann Druyan to continue his legacy. Ann and Seth MacFarlane were both critical of religion and prescience. They were a perfect match to put together the new Cosmos on Fox. His most famous book was Demon-Haunted World and the Pale Blue Dot.

This is something most people are unaware. You can find this in Martin Gardener's book.

---

[28] http://www.smithsonianmag.com/science-nature/why-carl-sagan-truly-irreplaceable-180949818/#tXkMYPyJ630wgTwl.99

In Quebec, Canada, Daniel L. Burnstein, a brilliant, leading Canadian physicist has created an alternative to the Standard Model of Particle Physics and possibly a viable Theory of Everything. Of course testing, research and falsification will bear out the results. It is called QGD. This is just an example of the creativity in the Scientific World.
http://www.quantumgeometrydynamics.com/blog/

For Great science writing, I highly recommend:

Faye Flam
http://www.philly.com/philly/columnists/faye_flam/ and Jennifer Ouellette, Cocktail Party Physics plus her other books.
http://blogs.scientificamerican.com/cocktail-party-physics/ and Ian O`Neil at
http://news.discovery.com/contributors/ian-oneill/

My favorite science Blog is Why Evolution is True, which is also the title of Jerry Coyne's fabulous book, on evolution. Jerry is a professor in the Department of Ecology and Evolution at the University of Chicago and he is also member of both the Committee on Genetics and the Committee on Evolutionary Biology.
http://whyevolutionistrue.wordpress.com/

Canada`s top skeptic and chemist is **Joe Schwarcz**, and I recommend his books Science, Sense and Nonsense and Is that a Fact? Joe is the Director of the "McGill Office for Chemistry and Society "at the McGill University in Montreal, Quebec, Canada. He takes on and confronts all pseudoscientific themes. http://www.mcgill.ca/oss/

Richard Dawkins is a British Ethnologist and evolutionary biologist who has written some incredible books on evolution and atheism. The two books I recommend are the "Greatest Show on Earth" and "The God Delusion". Richard and Lawrence Krauss are in a movie called "The Unbelievers". The film records the two of them touring and relaxing as part of their constant activity to bring science and the critical thinking process to the masses. You can watch as the two of them discuss Cosmology and Skepticism for large groups and their interactions with individuals that are skeptics and believers. I saw the film and it should be a priority for all aspiring Nerds.

Lisa Randall is a brilliant lecturer and prolific writer. She is a professor of science at Harvard. Her bio at Harvard states, "Professor Lisa Randall studies theoretical particle physics and cosmology at Harvard University. Her research connects theoretical insights to puzzles in our current understanding of the properties and interactions of matter." I enjoyed "Knocking at Heaven`s Door" and "Warped Passages".

Andrew Zimmerman Jones writes the Physics Guide for About.com, has studied and written about physics since 1991. He holds a Physics degree from Wabash College with honors & awards. If it were not for him, I would not have written this book. His book a Dummies Guide for String Theory captured my imagination, year ago when it first came out. It motivated me to take online course and read a multitude of other books. Thank you, Andrew.

David Darling and the Worlds of David Darling

http://www.daviddarling.info/index.htm will provide you with an extensive amount of cosmology and physics knowledge. He has links to rock and pop music also. He is a man of many hobbies and interests. He is an English astronomer and science writer. Visit his many worlds. You will not be disappointed. His latest book is the Rocket Man which is a history of Man in flight. http://www.amazon.com/The-Rocket-Man-David-Darling-ebook/dp/B00FH1RWTY

Brian Clegg is an English science author of topics from quantum physics to the biography of Roger Bacon. He wrote one of my favorite books on randomness called "Dice World". His latest book is Extra Sensory about ESP and telepathy from a science based evidence platform.

Brian Cox is a Royal Society physicist, who is well know as a great communicator of scientific thought, in the British Media. He continues to work at the Hadron Collider on particle physics. You can see many of his specials on BBC television or the BBC player, in North America. His latest projects have been the Science of Dr. Who and the Human Universe on the BBC. His latest books include "Why does E=mc2?" with Jeff Forshaw, and "The Quantum Universe", also with Jeff Forshaw. Brian Cox on Facebook is https://www.facebook.com/ProfessorBrianCox .

Michio Kaku: He is another great communicator and who is a Professor of Theoretical Physics at the City College of New York. He is an expert in Superstring Theory and a brilliant futurist. He was co-founder of String Field Theory. His latest books are Future of the Mind and Physics of the Future. Michio uses media and social media to popularize science. You can get more information at http://www.aps.org/careers/physicists/profiles/kaku.cfm .

Hank Green: He is an internet guru and a pop culture icon who doe the SciShow and videos for NASA and the magazine Mental Floss. http://www.youtube.com/user/scishow

Bill Nye, the science guy: Bill had an Emmy award winning show, on PBS, teaching science to young students. Bill is an atheist who believes that critical thinking and the literacy of science will make people leave the attraction of the supernatural and Creationism. He has appeared on talk shows and recorded videos for Walt Disney and even participated on Dancing with the Stars. He is a writer of science books, for children and continues to write articles for major magazines. He just made headlines when he debated Ken Ham of the Creationist Museum. http://www.youtube.com/watch?v=6mAyBwhiAJ8 and his credentials include John Hopkins and CEO of the Planetary Society.

If you want to get your weekly dose of science news try this link.
http://www.youtube.com/watch?v=B1tysX5G8To

Bad Astronomy by Phil Plait
will give the low down of what is reliable in science and what is not. Check out his Facebook site also.
http://www.badastronomy.com/index.html

Ethan Sigel
is an astrophysicist who has one of the best blogs I have ever read on Cosmology. The blog is called "It Starts with a Bang". He uses multimedia to teach his ideas and concepts. You must go there.
http://startswithabang.com/?page_id=4 and
http://scienceblogs.com/startswithabang/author/esiegel/

Sabine Hossenfelder
is another great scientist and blogger. Catch her at
http://backreaction.blogspot.ca/

I also want to state that Jennifer Ouellette has published the best book on Buffy and physics. It is called "The Physics of the Buffyverse". Buy it and enjoy.

Do want to see who are the most popular scientists? Here they are at http://lucretius21c.wordpress.com/2014/02/22/charting-physicists-and-astronomers-who-have-over-5000-twitter-followers/ Diagrams by lucretius21 c.

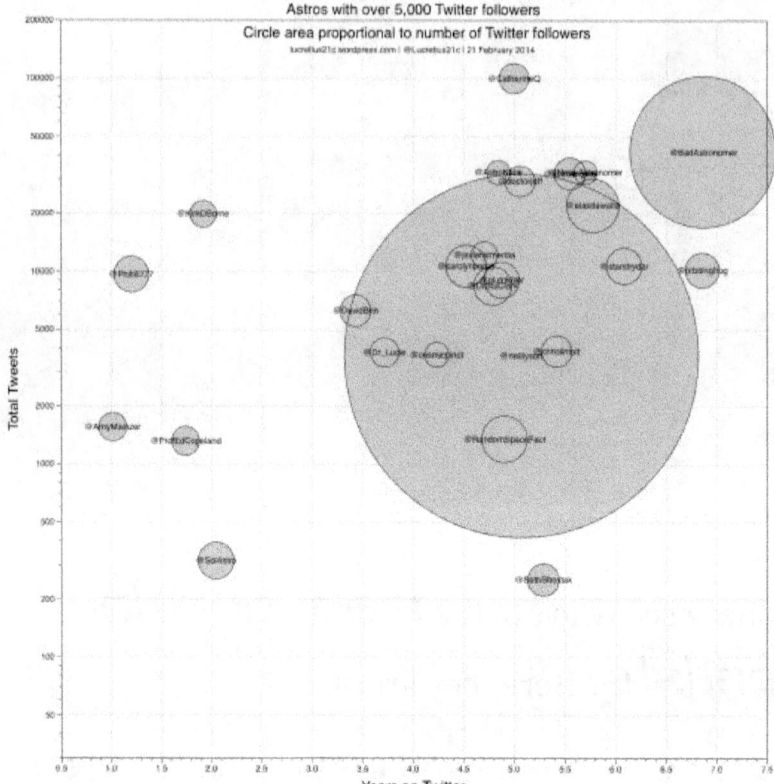

Click images for full-sized versions.

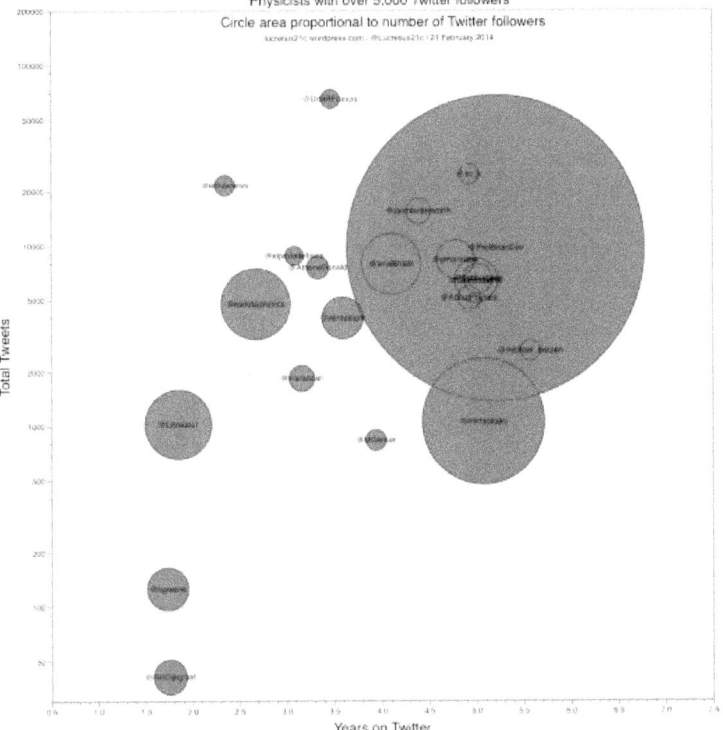

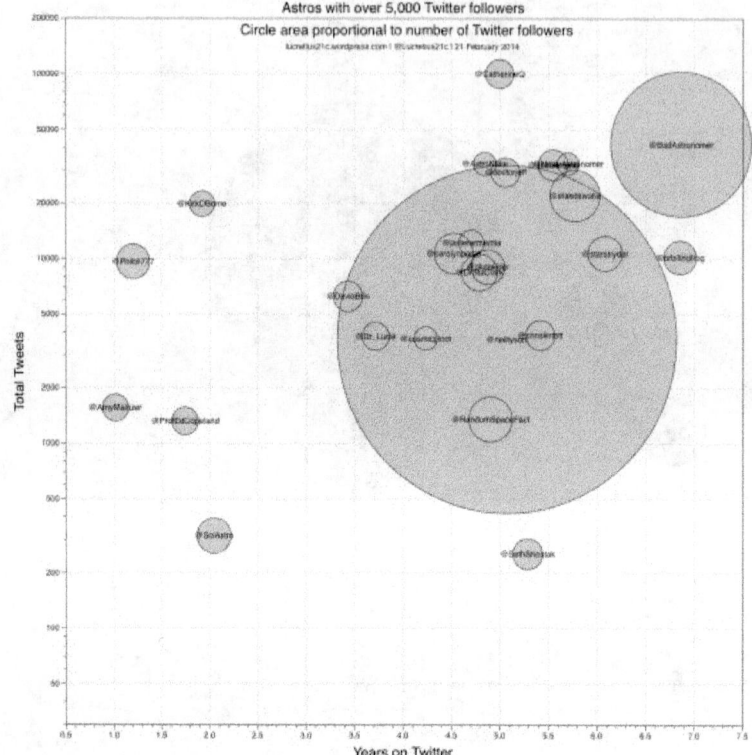

Posted on **2014-Feb-22 UTC** by Lucretius, ver. 21c.

Also see Physicists and Astronomers on Twitter.
http://lucretius21c.wordpress.com/physicists-on-twitter/

http://lucretius21c.wordpress.com/astronomers-on-twitter/

47. *Favorite Space site.*
www.space.com

48. *Favorite Science Website*
http://news.discovery.com/

49. *Coolest site and TV program for Nerds* http://www.g4tv.com/

50. *Best Nerd Humor Site*
http://www.theonion.com/

51. *Best Science Hip Artist*
GZA for his new CD called Dark Matter that takes the hip hop generation into learning about cosmology and physics through hip hop.
I thought I would end my book with my favorite comic and musical band, in a fine Nerd fashion.

Torch, Invisible Woman and Mr. Fantastic.

THIS is METALLICA, the band that has been my favorite band for the last twenty years. Also, I think that is **SLIPKNOT** is the metal band of the future.

Public Domain
http://upload.wikimedia.org/wikipedia/commons/c/c6/Metaltown_2009%2C_Slipknot.jpg/

Finally, do you want to hear over 1700 lectures? TED TALKS.

https://www.ted.com/talks/browse

Source Flickr: Metallica_46

Author Tony

http://www.youtube.com/watch?v=maLFlBImiQY
Ride the Lightening in Rio 2011.

    Finally, fear, not because at the end of the Nerd Universe is a story that the Fortress of Solitude is real. Frank Wilczek has created, on paper, based on Quantum Mechanics, a crystal computer that can survive the death of the universe. This computer would contain enormous amounts of information, lodged in the crystals, about our universe after it has succumbed to entropy and is long gone. Sound familiar. It is called the Fortress of Solitude, where crystal memory embodies everything there is to know about Krypton and its universe, though it has been annihilated. All we need now is a real Superman.

ⓘ ⓒ Some rights reserved by deryckh

http://www.newscientist.com/article/dn22028-computer-that-could-outlive-the-universe-a-step-closer.html

Now, congratulations you have learned many of the ideas related to physics and cosmology. You are a science nerd.

If you enjoyed this book please tell others about it and also leave a positive review on Amazon. Thank you.
https://www.amazon.com/gp/help/customer/display.html?nodeId=200791000 **or go to the book.**

Bibliography

Gubser, Steven Scott. *The Little Book of String Theory*. Princeton: Princeton UP, 2010. Print

Gribbin, John. *Schrodinger`s Kittens and the Search for Reality*. Boston: Little Brown and 1995. Print.

Jones, Andrew Zimmerman., and Daniel Robbins. *String Theory for Dummies*. Hoboken, NJ: Wiley, 2010. Print.

Orzel, Chad. *How to Teach Physics to Your Dog*. New York: Scribner, 2009. Print.

Pickover, Clifford A. *The Physics Book: From the Big Bang to Quantum Resurrection, 250 Milestones in the History of Physics*. New York: Sterling Pub., 2011. Print.

Al-Khalili, Jim. *Quantum: A Guide for the Perplexed*. London: Phoenix, 2012. Print.

Thomas, Andrew. *Hidden in Plain Sight*. Swansea: A. Thomas, 2012. Print.

`Rae, Alastair I. M. *Quantum Physics: A Beginner's Guide*. Oxford: Oneworld, 2005. Print.

Cox, Brian, and J. R. Forshaw. *The Quantum Universe: Everything That Can Happen Does Happens*. London: Allen Lane, 2011. Print.

Carroll, Sean M. *The Particle at the End of the Universe: How the Hunt for the Higgs Boson Leads Us to the Edge of a New World*. New York: Dutton, 2012. Print.

Gribbin, John. *In Search of Schrödinger's Cat: Quantum Physics and Reality*. Toronto: Bantam, 1984. Print.

Ford, Kenneth William. *101 Quantum Questions: What You Need to Know about the World You Can't See*. Cambridge, MA: Harvard UP, 2011. Print.

Stenger, Victor J. *God and the Folly of Faith: The Incompatibility of Science and Religion*. Amherst, NY: Prometheus, 2012. Print.

Stenger, Victor J. *God and the Atom*. New York: Prometheus, 2012. Print.

Clegg, Brian. *Dice World: Science and Life in a Random Universe*. London: Icon, 2013. Print.

Gribbin, John. *Erwin Schrödinger and the Quantum Revolution*. Hoboken, NJ: Wiley, 2013. Print.

Calle, Carlos I. *Einstein for Dummies*. Hoboken, NJ: Wiley Pub., 2005. Print.

Baggott, J. E. *The Quantum Story: A History in 40 Moments*. Oxford [England: Oxford UP, 2011. Print.

Kaku, Michio. *Parallel Worlds: A Journey through Creation, Higher Dimensions, and the Future of the Cosmos*. New York: Doubleday, 2005. Print.

Ford, Kenneth William. *The Quantum World: Quantum Physics for Everyone*. Cambridge, MA: Harvard UP, 2004. Print.

Baker, Joanne. *50 Physics Ideas: You Really Need to Know*. London: Quercus, 2007. Print.

Greene, B. *The Hidden Reality: Parallel Universes and the Deep Laws of the Cosmos*. New York: Alfred A. Knopf, 2011. Print.

Krauss, Lawrence Maxwell. *A Universe from Nothing: Why There Is Something Rather than Nothing*. New York: Free, 2012. Print.

Carroll, Sean M. *From Eternity to Here: The Quest for the Ultimate Theory of Time*. New York: Dutton, 2010. Print.

Lloyd, Seth. *Programming the Universe: A Quantum Computer Scientist Takes on the Cosmos*. New York: Knopf, 2006. Print.

Hawking, Stephen, and Leonard Mlodinow. *The Grand Design*. New York: Bantam, 2010. Print.

Mlodinow, Leonard. *The Drunkard's Walk: How Randomness Rules Our Lives*. New York: Pantheon, 2008. Print.

The Worlds of David Darling. David Darling, n.d. Web. <http://www.daviddarling.info/index.html>.

Recommended Books

Gribbin, John. *Schrodinger`s Kittens and the Search for Reality*
http://www.amazon.com/Schrodingers-Kittens-Reality-Gribbin-Paperback/dp/B00GSCZBVW/ref=sr_1_4?ie=UTF8&qid=1399247220&sr=84&keywords=Schrodinger%60s+Kittens+and+the+Search+for+Reality

Carroll, Sean M. *From Eternity to Here: The Quest for the Ultimate Theory of Time*
http://www.amazon.com/From-Eternity-Here-Ultimate-Theory/dp/1851688951/ref=sr_1_2?ie=UTF8&qid=1399247443&sr=82&keywords=Carroll%2C+Sean+M.+From+Eternity+to+Here%3A+The+Quest+for+the+Ultimate+Theory+of+Time

Krauss, Lawrence Maxwell. *A Universe from Nothing: Why There Is Something Rather than Nothing.*
http://www.amazon.com/Universe-Nothing-There-Something-Ratherebook/dp/B004T4KQJS/ref=sr_1_1?ie=UTF8&qid=1399247560&sr=81&keywords=Krauss%2C+Lawrence+Maxwell.+A+Universe+from+Nothing%3A+Why+There+Is+Something+Rather+than+Nothing.

Gribbin, John. *Erwin Schrödinger and the Quantum Revolution.*
http://www.amazon.com/ERWIN-SCHRODINGER-AND-QUANTUMREVOLUTION/dp/B00IFGGL86/ref=sr_1_7?ie=UTF8&qid=1399247717&sr=8-7&keywords=Gribbin%2C+John.+Erwin

God and the Atom **by Stenger, Victor J.**

http://www.amazon.com/God-Atom-Victor-J-Stenger-ebook/dp/B00BH0VP0Q/ref=sr_1_4?s=books&ie=UTF8&qid=1399247753&sr=1-4&keywords=victor+stenger

Carroll, Sean M. *The Particle at the End of the Universe*
http://www.amazon.ca/Particle-End-Universe-Higgs-Boson/dp/0525953590/ref=sr_1_1?s=books&ie=UTF8&qid=1408045708&sr=11&keywords=particle+at+the+end+of+the+univere

Lloyd, Seth. *Programming the Universe: A Quantum Computer Scientist Takes on the Cosmos.*
http://www.amazon.com/Programming-Universe-Quantum-ComputerScientistebook/dp/B000GCFBP6/ref=sr_1_1?s=books&ie=UTF8&qid=1399247958&sr=1-1&keywords=Lloyd%2C+Seth.+Programming+the+Universe%3A+A+Quantum+Computer+Scientist+Takes+on+the+Cosmos

Hawking, Stephen, and Leonard Mlodinow. *The Grand Design*
http://www.amazon.com/Grand-Design-Stephen-Hawking-ebook/dp/B003TXSF5C/ref=sr_1_fkmr0_1?s=books&ie=UTF8&qid=1399248048&sr=1-fkmr0&keywords=Hawking%2C+Stephen%2C+and+Leonard+Mlodinow.+The+Grand+Designntist+Takes+on+the+Cosmos.

The Hidden Reality: Parallel Universes and the Deep Laws of the Cosmos by Brian Greene

http://www.amazon.com/Hidden-Reality-Parallel-Universes-Cosmos/dp/0307265633/ref=sr_1_1?s=books&ie=UTF8&qid=1399248140&sr=1-1&keywords=brian+greene

What as a Nerd or Geek

From Talent Culture:
http://www.talentculture.com/career-strategy/dare-to-be-different-5-reasons-geeks-get-great-jobs/

"No matter what your interests may be, if you're a geek, you tend to be creative and experimental. Since you don't learn via traditional methods or work via traditional processes, you tend to find ways to teach yourself."

Geek Pride Day Survey, May 25, 2013

http://www.modis.com/it-insights/press-room/survey-downloads/modis-geek-pride-survey-2013-media-deck.pdf

Intelligence, nerd clothing, video games, comic books, science fiction, computers, general technology and glasses are the inner nerd and geek.

ACKNOWLEDGEMENTS

I would like to thank and acknowledge all the individuals who continually motivated me, assisted and advised me, in the creation of this book. They include, Gary Herman, my close friend from IBM Canada, brilliant sales, technological and science person, who especially put a lot time into my work. This book could not have been written without him. Gary you are the best. Nicola Ormerod, for her support while motivating me to write this book. She is CEO and owner of Karabeth Publishing. Nicola thanks for answering all those questions that I constantly bombarded you with on Facebook. Steve Guest, a good friend, and CEO of S.Graphics. He has an amazing graphics company; thank you for your support and inspiration. Steve led me through the darkness of organizing, conjuring the content and designing my brilliant cover. Many have said they would buy the book for his cover. I also want to show my appreciation to Louise Ouimet , librarian teacher at Daystrom Public School in Toronto. She provided me with the knowledge and motivation to write this book. Louise is an amazing teacher. Also, aspiring author Keisha Martin for her suggestions and support throughout the writing process. I want to thank my brother David for the encouragement he gave me. I want to thank the support of the people who endorsed me without a published copy. I am indebted to Gail Moss Baymiller, Guy P. Harrison, Daniel Burnstein, David Darling, Tommy Rodriguez and Carmelo Nanfara. There are no words to describe what good people you are for taking the time to read it and comment. Finally, to my wife, Jill and my daughter Alana and son Michael, who were instrumental in supporting me; *thank you.* Also, thank you to everyone who works hard to try and be scientifically literate. That is difficult in an age of conspiracy theories and alternative medicine and homeopathy. Thank you all. Ian